THE
INTIMATE
ANIMAL

THE INTIMATE ANIMAL

The Science of Love, Fidelity and Connection

DR JUSTIN GARCIA

PENGUIN LIFE

AN IMPRINT OF

PENGUIN BOOKS

PENGUIN LIFE

UK | USA | Canada | Ireland | Australia
India | New Zealand | South Africa

Penguin Life is part of the Penguin Random House group of companies
whose addresses can be found at global.penguinrandomhouse.com

Penguin Random House UK,
One Embassy Gardens, 8 Viaduct Gardens, London SW11 7BW

penguin.co.uk

First published in the United States of America by Little, Brown Spark an imprint of Little,
Brown and Company, a division of Hachette Book Group Inc 2026
First published in Great Britain by Penguin Life 2026
001

Copyright © Justin R. Garcia, 2026

The moral right of the author has been asserted

Penguin Random House values and supports copyright.
Copyright fuels creativity, encourages diverse voices, promotes freedom
of expression and supports a vibrant culture. Thank you for purchasing
an authorized edition of this book and for respecting intellectual property
laws by not reproducing, scanning or distributing any part of it by any
means without permission. You are supporting authors and enabling
Penguin Random House to continue to publish books for everyone.
No part of this book may be used or reproduced in any manner for the
purpose of training artificial intelligence technologies or systems. In accordance
with Article 4(3) of the DSM Directive 2019/790, Penguin Random House
expressly reserves this work from the text and data mining exception

Printed and bound in Great Britain by Clays Ltd, Elcograf S.p.A.

The authorized representative in the EEA is Penguin Random House Ireland,
Morrison Chambers, 32 Nassau Street, Dublin D02 YH68

A CIP catalogue record for this book is available from the British Library

ISBN: 978-0-241-42593-0

Penguin Random House is committed to a sustainable future
for our business, our readers and our planet. This book is made from
Forest Stewardship Council® certified paper.

*To my kin, for continuously showing me the spectacular
power of love in all its dimensions:
My wife, Michelle; my parents, Helen and Mike;
my family Jen, Dave, Emily, and Alexis.*

CONTENTS

Introduction: *The White Whale* — 3

Chapter One: Need — 17

Chapter Two: Crave — 29

Chapter Three: Search — 51

Chapter Four: Date — 73

Chapter Five: Mate — 95

Chapter Six: Nest — 117

Chapter Seven: Stray — 135

Chapter Eight: Break — 155

Chapter Nine: Care — 169

Chapter Ten: Love Again — 183

Conclusion: *Living and Dying for Love* — 197

Acknowledgments — *203*
Notes — *206*
Index — *257*

THE INTIMATE ANIMAL

INTRODUCTION

THE WHITE WHALE

Pahrump, Nevada, is a tiny town tucked unobtrusively into the desert about an hour outside of Las Vegas. It has fewer residents than Indiana University Bloomington, where my lab is based, has students. But it receives plenty of visitors on a daily basis, thanks to its two legal brothels.

I was traveling to Pahrump in the name of science. Over lunch at a conference a few months earlier, my colleagues Michelle Escasa-Dorne and Peter Gray and I had hatched a plan to collect some preliminary observational data on the operations of a legal brothel, both the daily lives of its employees and the habits and preferences of clients.[1] We chose the brothel with the better Yelp reviews for our initial field research, and before I knew it, I was heading west to join my fellow scientists in the middle of the desert.

Our car pulled up to the brothel as the sun was setting, the neon lights rising from the oasis of the desert floor. The building, which looked like a converted one-story motel, had two entrances: one to a restaurant and bar, the other to GIRLS! GIRLS! GIRLS! We chose door number two, showed our IDs to a cheerful middle-aged woman in business-casual attire who introduced herself as the house manager, and filed into a long, tall-ceilinged vestibule. Tile floors, all the furniture recently polished. Air smelling of lilac and Lysol. It was dark but inviting in a luxurious way.

The manager rang a bell, summoning our guide, who entered behind us, drawing our gaze to a large poster propped on an easel, which turned out to be a menu of sexual acts. Raquel was a former flight attendant turned sex worker with bright yellow hair and a Minnie Mouse voice. Married with two children, she began working at the brothel after losing her job during the economic crash of 2008. She was a gracious guide, affable in her understanding that we were on a scientific scouting mission of sorts.

My colleagues and I arrayed ourselves around the easel, and Raquel took us through the options. "Our clients come from a huge range of backgrounds," she explained. "We get bachelor and bachelorette parties from Vegas, as well as guys and gals who have hit it big at the craps table and come on over to celebrate."

The list of sex acts on the menu seemed fairly straightforward. I scanned the less expensive options, which included everything from sexual intercourse and shower parties to breast massage and drag dress-up, before my attention was drawn to the "special" section of the menu — role-playing, lingerie, couples' threesomes, and more. That was when I spotted the most expensive item on the menu.

"What's the White Whale?" I asked.

Starting at a base price of $20,000 (each worker set her own rates), it was clearly not for the faint of heart.

"Oh, that's the full Girlfriend Experience. A great choice for someone who wants something... personal," she said, adding that it was "a house favorite among high rollers."

"Sex isn't *necessarily* a part of it," she told us. "But you'll get a hell of a cuddle."

We were struck silent by this revelation. Here was the most expensive sex act money could buy in a legal brothel, and it didn't necessarily involve sex.

What people were buying was *intimacy*.

THE PARADOX

Intimacy is a broad term and a nebulous scientific concept. We use the word to describe a wide range of connections: the bond shared by a romantic couple, the unconditional love between parent and child, even the trust and support of close friends. But what do we really mean when we talk about intimacy?

In the abstract, intimacy is the pleasurable and comforting feeling associated with any close connection that grows between humans in a huge variety of contexts.[2] In practice, it's making eye contact across the table at a dinner party and knowing exactly what the other person is thinking; it's feeling safe enough to lower your emotional armor and expose your deepest insecurities; it's someone else sensing what you need, even before you know it yourself. In other words, it's the experience of closeness, of feeling and being seen, heard, and known.

Intimacy is the foundation of every successful romantic relationship — whether it be between two people of different genders or the same gender; whether it's among the young experiencing first love or seniors hoping to recast their closest connections; whether the relationship is monogamous, nonmonogamous, or polyamorous. There can also be intimacy in other types of relationships: with friends, with family, even with co-workers. Intimacy is at the very core of the human condition, explaining so many of our best and worst behaviors.

Yet so few of us understand this essential drive — how it has impacted the evolution of our species, how it lives just under the surface of our desires, or how to harness it. We might not even recognize the need for intimacy as a biological *drive,* perhaps because it lives in the shadow of that *other* primal urge: our sex drive.

We tend to think of our sex drive as the most powerful evolutionary motivator of modern relationships.* For decades, evolutionary

---

* Sexual behavior in humans involves both instinctual fixed action patterns and learned behaviors. I use the term *sex drive* here to reference the internal, motivational state that pushes an organism to engage in specific behaviors in service of survival and reproduction. Some scholars use the psychoanalytic term *libido* in place of *sex drive*. In *Come as You Are* Emily Nagoski thoughtfully critiques *sex drive* as a term and concept that overly emphasizes sex as a need and that fails to address how sexual desire may operate as an incentive motivational system, often responsive to stimuli rather than spontaneous. While it is important to remember that most adaptations related to our intimate lives respond to ecological and social context, I use the term *sex drive* because sex behavior is an evolutionary need in the sense that without sex and sexual reproduction our genes and species do not survive to the next generation. However, the presence of a drive does not mean people are entitled to sexual activity with others and in no way justifies nonconsensual behaviors.

biologists have focused on our sex drive as an evolutionary adaptation designed to motivate mating and reproductive behaviors necessary for the species's survival.[3] But in focusing on sex and reproduction as the primary motive for romantic relationships, we have neglected a complementary core truth: that our motivation for intimacy and love is distinct from our sex drive. And it is also much stronger than we have been led to believe.

As my collaborator and friend the late renowned anthropologist Helen Fisher would say, "Everywhere in the world people pine for love, live for love, kill for love, and die for love."[4]

As an evolutionary biologist, sexologist, and university professor, I have devoted most of my professional life to researching romantic and sexual relationships and the many variations of human sexual behavior and intimacy. The range of wants and needs I am fortunate enough to study is bountiful, beautiful, and sometimes mind-boggling. This diversity in our intimate lives reveals something profound about the complexity of the human experience.

At the intersection of love and sex lies a fundamentally intractable evolutionary paradox: Humans are wired to be socially monogamous — that is, we have a remarkable capacity and desire to form intense pair bonds with other humans, usually one at a time, and sometimes lifelong — but we are not necessarily wired to be sexually monogamous.

What this means is that our sexual impulses are often in *direct opposition* to our existential need for love and intimacy.

My research, in the broadest sense, seeks to provide scientific context to what poets have pondered for centuries: Our desire for sex and our need for connection are powerful forces that underwrite so many of the peaks and valleys of our romantic and sexual relationships. These forces are baked into our deep evolutionary story as a highly social mammal and are intertwined with our biology, psychology, and cultural systems.

When these two evolved drives — for sex and for intimacy — are in sync, we feel the kind of love and passion that poets dream of: all-consuming in its power and pleasure. The highs can be magically high.

But the lows can also be painfully low. When our desires for intimacy and sex are at odds, we often find ourselves unhappy; we may pick a partner who satisfies us sexually but not emotionally, or vice versa, leaving us disappointed, heartbroken, unfulfilled. Many of the mistakes or bad choices that we make in relationships, both big and small — dating people who are wrong for us, drifting apart from a long-term partner, shattering the trust we share with another person — arise from this fundamental tension between our evolved desire for sex and our evolved need for intimacy.

So, we're left with a whale of a question: Can we reconcile our competing desires to enjoy a deeper and more satisfying form of romantic love? I think we can. But it requires a new understanding of the evolutionary processes that continue to shape our romantic and sexual lives.

In the chapters ahead, we'll explore how to harness this new understanding of intimacy to sustain the peaks and smooth out some of the inevitable valleys in our romantic adventures.

SCIENTISTS OF THE HEART

I was five years old when I dispensed my first piece of advice about love. I had just gotten a close-up look at relationship behavior in a young adult male, and I did not like what I saw.

"Your boyfriend," I announced to my then eighteen-year-old cousin, Jen, "is a *jerk*."

It was an August afternoon on the South Shore of Long Island, New York. The humid air, slightly fogged with sea and salt, had made everything a little bit damp. I was crouched down on the driveway of Jen's house, two tiny fists stacked on the handle of my magnifying glass and my face millimeters away from a line of ants streaming along a crack in the concrete. Close observation of the species required care and patience. You did not train the magnifier on individual ants. Hot summer sun, I had discovered, could turn the instrument into a death ray.

I was a little bit in love with ants. My mother had bought me an insect explorer kit, and I couldn't get enough of it. From an encyclopedia on a low shelf at home, I had amassed an abundance of amazing ant facts. (A personal favorite: Ants live on every continent except *Ant*arctica.)

Observing these fascinating creatures "in the wild" was thrilling. I'd been running a highly unscientific experiment on how driveway ants navigate melted popsicles when Jen and Chad drove up in his Jeep. Jen and I were both only children with no other cousins, and the familial connection we intuitively shared was something I wouldn't understand until I was old enough to study kinship bonds — which, in humans, are not unlike the kinship behaviors in the ants I was watching.

Jen introduced me to Chad, a young surfer dude who exuded a cool, casual vibe. I took note of how he rolled his eyes as she knelt down to ask me about my ant friends. Jen listened patiently as I explained that an ant can lift more than twenty times its own body weight (that's like a child lifting a car!) before she led Chad inside to say hello to her mother and stepfather. I followed, magnifying glass in hand.

At five years old I didn't yet have the words or tools to articulate why I was suspicious of Chad. Many years later, I would understand why the qualities that make for an exciting summer romance aren't necessarily the same that make for a devoted long-term partner, but at the time, it was simply clear from the moment I met Chad that he was a *bad* ant — and I wished for a surfer-sized magnifying glass so I could light him up.

In the end, my intuition wasn't wrong. As it turned out, Chad *was* a jerk, less interested in the things and people his partner cared about than in his own surf gear, and less concerned with interpersonal commitment than with his commitment to partying on the beach. But I didn't know any of that then. All I knew was what I could observe in that moment: He wasn't kind to me, the way Jen was. The fact that she cared about me didn't seem to matter to him.

My understanding of exactly *why* a jerk like Chad wasn't a suitable partner for Jen — and why she had been attracted to him anyway, at least for a brief period of time — would come many years later, as I began exploring the intersections between evolutionary biology and human social and intimate behavior. When I moved from New York to Bloomington, Indiana, in 2011 to begin what was supposed to be a two-year postdoctoral research fellowship, I never imagined that I would spend the next decade-plus intensely researching people's sexual and romantic lives through an evolutionary lens, eventually publishing dozens of papers and becoming the eighth (and youngest) executive director of the Kinsey Institute since it was established in 1947.

At the time, I didn't think of myself as a sex researcher in the traditional sense. Rather than being a sexologist who used evolutionary theory in their work, I saw myself as an evolutionary biologist who had wandered into studying sexuality as an extension of my interests in social behavior and mating. When this incredible fellowship opportunity presented itself, I thought, what better place to shore up that side of my training than at the mecca of sex research?

Widely considered the grandfather of Western sexual science, Alfred C. Kinsey was a biologist and professor of zoology who had devoted much of his early career to creating a detailed taxonomy of gall wasps.[5] It wasn't until 1938, two decades into his tenure at Indiana University Bloomington, when he was asked by the university to co-teach a new class on marriage and sexual hygiene (for married students only, of course), that he trained his proverbial magnifying glass on the science of sex.[6] Kinsey was tasked with teaching the physiology component of the course, the thinking presumably being that because he was a zoologist with an interest in insects (a decidedly unsexy group of species), his approach to teaching sex and reproduction would be relatively uncontroversial.

They thought wrong.

What ended up happening instead in many ways represents the very best of what university life is supposed to be about. You teach a course, and the students' questions enrich the way you think about the topics; in this instance, there were so many questions about human sexuality that Kinsey couldn't answer because *no one* had the answers. Realizing there was virtually no scientific work on this subject, Kinsey and his colleagues began systematically compiling sexual case histories on a massive scale. With over 18,000 interviews of Americans from every walk of life conducted, it remains, to this day, one of the richest sexological studies ever conducted — and would ultimately become the basis of two groundbreaking and at times controversial books, *Sexual Behavior in the Human Male* (1948) and *Sexual Behavior in the Human Female* (1953).[7]

These two seminal works, also known as the Kinsey Reports, gained international attention for shattering widespread myths and proving unorthodox theories such as the previously unthinkable idea that women, too, are sexual beings — a revelation that played no small part in liberating female sexuality.[8] The study also revealed that many of the sexual behaviors that were illegal or considered "deviant" were actually quite common in every sector of American society. This research soon led to the development of the Kinsey Scale, which describes sexual orientation as a continuum, replacing the then culturally recognized binary of heterosexual versus homosexual.[9] In short, Kinsey's work had a profound impact on American culture that reverberates to this day.[10]

I am the first evolutionary biologist at the helm of the Institute since Alfred Kinsey himself. It's a prodigious legacy, and a role that I approach with humility. Along with a team of multidisciplinary researchers and students, I spend my days investigating the science behind the vagaries and varieties of the human heart and communicating our findings to other academics and the world. What I've learned is that the assumptions many of us have about relationships don't hold up under the magnifying glass of the biological and behavioral sciences.

For example, the idea that passion inevitably dies when we start to nest is a myth. Though the trope of the sexless middle-aged couple might be a reliable source of laughs for network sitcoms and anniversary toasts, the science shows that companionate love can be just as fiery as first-blush crushes.[11]

Other assumptions that don't hold up to scrutiny include the idea that breakups are harder on women than on men, and that women are always the first in a relationship to say "I love you." Despite what all the classic rom-coms would have you believe, men actually take longer to grieve relationships, on average, than women do, and they also — on average — say "I love you" first.

Then there's the myth that the young are more likely to value looks than older people. In fact, both teenagers and seniors prioritize the same quality in someone with whom they want to start a relationship, and it's not physical attractiveness — it's how trustworthy the person is. We may primarily chase beauty for flings, at any age, but when it comes to relationships, looks take a backseat to what we all want in a long-term partner: someone we can trust and confide in.

Understanding our evolved biology sheds light on the innate motivational systems that explain whom and how we love. Without this understanding, it's impossible to make sense of what we want and need, sexually and emotionally, from our partners — or what our partners, in turn, need and want from us.

Historically, love hasn't been taken all that seriously as a field of scientific inquiry. Psychoanalyst Sigmund Freud, for instance, had lots to say about sex but was a bit stumped by love.[12] Even Kinsey thought love too squishy to measure, too difficult to operationalize, and therefore didn't include it in his studies of human sexuality.[13]

My inspiring team and I continue to explore the biological and cultural landscape of human sexuality and relationships. The mission of the Institute has grown from its early groundbreaking work uncovering the diversity of human sexual behavior to include scientific examinations of the interconnected role of sexuality and relationships. Today, we study such multifaceted ideas as the social context of sexual and romantic experiences and the intersections of sex and sexuality with human love. That does *not* mean that we only recognize sex that occurs in the context of committed relationships — we are equally interested in the sexual experiences (and lack thereof) that occur with casual partners, with multiple partners, in the absence of a partner, and/or with someone to whom we are pair-bonded and deeply in love. In fact, it's precisely this rich variation that we explore in many of our research studies.

Though the comparative research literature on relationships among LGBTQ+ populations has, until now, been woefully lacking, the work we are doing at the Kinsey Institute is dedicated to widening the lens of scientific research to examine the full spectrum of gender and sexuality and diverse pursuits of human intimacy.[14] If the Institute's work in the early days was a kind of Newtonian physics, establishing basic laws and assumptions, we're now in the quantum age, pushing new boundaries and exploring new frontiers in human sexuality — including all the ways in which relationships can look, feel, and be experienced differently across sexual orientations and gender identities. I like to think that Kinsey would be proud.

In addition to my work at the Institute, I am scientific advisor to Match, known for the popular online dating site/app Match.com: a role that allowed me to team up with one of my mentors and collaborators, anthropologist Helen Fisher, to take on an enormous task of trying to understand the attitudes and behaviors of the rising demographic of single adults in an ever-changing dating landscape.[15]

Since launching our first survey in 2010, the goal of this ongoing Singles in America study has been to provide new insight into the complexities of modern mating. What we have discovered is that, in some ways, a brothel in the Nevada desert isn't all that different from everyday bedrooms around the world. While the contexts may be different, these interactions are beholden to the same human impulses. In other words, our research is tapping into something that is surprisingly high up on the hierarchy of human needs: sex that *invokes intimate connection* with another person, even if that sense of connection is only fleeting or an illusion.

Our research is being used to illuminate the most human of interactions: love, sex, and intimacy. Applying bench-to-bedside research with rigorous scientific heft to the study of love and sex, we hope to better understand the spectrum of human intimate experiences: why we want whom we want; why we stay or stray; the real reason sex is important to relationship maintenance; why our species has evolved to live and die for love.

Whether you are still searching for that special someone, considering taking a relationship to the next level, or looking for ways to keep the romance of a stable, long-term relationship alive, my goal is to equip you with a deeper understanding of how the human animal navigates the intimate relationships that define our lives.

WHAT'S GOOD FOR YOU, ME, AND US

My kindergarten birthday party was held at a spectacular hundred-year-old carousel in Forest Park, Queens. Jen — who had laughed so wholeheartedly when I told her Chad was a jerk — remained my favorite (and only) cousin, so of course she came to my party. I remember basking in the light of her attention as she hoisted me up onto a white, snarling mustang frozen in mid-stride, a horse I both coveted and feared. She had brought her new boyfriend, Dave, and asked if I wanted to meet him.

Of course I did.

After the ride, she led me, flush with adventure, past the face-painting clown to meet him. Dave and I spoke of horses, then of sharks — by then, I had moved on from ants to more complex animals.

I was enamored. So was Jen. I approved of her new boyfriend. It seems the rest of the family did as well; Jen and Dave eventually married and began a life together. I was the youngest member of their wedding party.

When I got a little older, I started to wonder about the odds of obtaining that kind of connection in my own romantic relationships, that cocktail of friendship and passionate desire they openly shared, that intoxicating mix of something both exciting and safe. I wondered about the universality of what I had observed between the two of them: Do we all want to find and create a life with someone who looks at us the way Dave looked at Jen and Jen looked at Dave?

I believe that we do. But part of the complexity of modern mating is that there isn't a one-size-fits-all approach to achieving this holy grail. Yes, we can read dozens of books offering advice on how to navigate relationship conflict, and many of them will be extremely helpful, but the challenge is that human beings have significant individual differences in how we approach and understand our relationships. It's why, after eight decades of research, we don't have a simple three-page how-to guide when it comes to human intimacy. There's too much individual difference. That variability is messy, but it's also what makes relationships so exciting and self-expanding. We may never fully crack the universal code, but as individuals and as couples, we can take pieces of the science to make sense of the world around us and use what works to enrich our relationships.

Like any enduring work of art, we must craft our most beautiful relationships. And we do that by *investing* in each other, by finding time to be together, and by showing vulnerability and trust. No relationship is perfect, and very few people would say there's nothing they would change about their partners, but compromise is how we honor the complexities of building a life with another human. One of my favorite pieces of relationship advice once shared by a therapist colleague is to prioritize these three questions: *What's good for me? What's good for you? What's good for us?*

Introduction

Romantic relationships are simultaneously messy and beautiful, but when we decide that's what we want, the next intimate impulse is often to turn the relationship into something bigger than itself, to make a home and create a family within all the complexity and excitement and confusion of the modern world.

Chapter One

NEED

When people learn that I study intimacy and human sexuality for a living, they tend to tell me things. Personal things. Private things.

Not long ago, on a flight to Vancouver, I struck up a conversation with the woman sitting next to me. Ginny had warm brown eyes and that disarming midwestern charm that makes you feel as though you'd be invited to the family barbecue. As our chatting evolved from casual exchanges about where we were headed and what we were reading to a more personal conversation about our lives and respective careers, she flagged down the flight attendant and ordered two tiny bottles of chardonnay.

As we sipped and talked, I learned that Ginny had struggled with depression since her mother was diagnosed with Alzheimer's disease a few years earlier. At the suggestion of her therapist, she had started taking dance classes in the evenings after work.

"Honestly, it felt so awkward at first, but there was something about just letting go and allowing myself to be vulnerable in front of all those new people that was freeing. It opened me up and pulled me out of the rut I was in. I feel more alive now than I have in a long time."

When she mentioned that she and her husband, Matthew, were coming up on their tenth wedding anniversary, I congratulated her on the milestone. A look I've come to recognize as the prelude to an intimate secret passed across her face. "I love my husband," she sighed, "but more and more it feels like we're living separate lives. I can't even remember the last time we... you know."

This kind of airplane exchange is not unusual for me. Over the course of nearly two decades researching people's intimate lives, I've met a lot of people and I've heard a lot of things. The tension that almost always lies at the heart of their stories is precisely what makes human intimacy so fascinating to study, and so complicated to navigate.

Love in all its forms is a constellation of feelings, experiences, and drives — for emotional connection, for sexual pleasure, for reproduction, for anchoring our sense of self in our social world.[1] Humans evolved this tendency to form intense romantic bonds — *pair bonds* in the language of science — because it allowed us to master uncertainty, to not just survive but thrive in a world that is both rife with danger and filled with boundless opportunity.

We know from some of the best research studies on pair bonds — how they form, how they dissolve, how they are associated with nest building, territory defense, and survival — that this adaptation has played a critical role in the proliferation of a wide variety of species across the natural world.[2] What I find endlessly fascinating about the evolution of human pair bonds, however, is that it tells a meta-story about our species that goes far beyond reproduction and sexual selection. The instinctual drive for human connection, coupled with our desire for sexual experience, is what motivates us to seek out romantic partnerships.

I find myself at the heart of sex research during a pivotal moment in our sexual evolution. We live in a time of totally redefined gender roles and relationship norms, filled with clicks and swipes, and courtship not bound by geography or tradition. We have seemingly infinite access to potential mates, seemingly infinite potential for being swept off our feet, seemingly infinite technological capacity for chatting, flirting, and loving globally.

But the research shows that we find ourselves stuck. Rates of depression and loneliness are on the rise, even among people in relationships. People of all ages, but especially young adults, are reporting burnout with today's dating norms. There are proportionally more single adults than ever before, and our mating patterns are in flux around the globe.[3]

In the United States alone, close to 40 percent of the adult population is single. That's well over 120 million adults, moving in and out of romantic and sexual relationships across the course of a life.[4] There is almost no other society we know of in the cross-cultural literature where so many adults have been single at a given time. There is evidence that something like this is happening in contemporary Japan and some other industrialized nations, suggesting we may be on the shoreline ahead of a global singlehood wave. But for now, the demographic pattern is still atypical.

Why is this? What's going on? One dominant theory is that it's a pattern that's been taking shape since the aftermath of World War II, when greater industrialization meant that women started having more autonomy and faced less social and economic pressure to pair up. But I would posit that it's also because the very definition of relationships is changing. Fifty years ago, a relationship implied exclusivity; today, it does not. Just ask any nineteen-year-old, for whom an entirely new vocabulary of terms — "just talking," "hooking up," "hanging out" — has replaced the old binary of "single" versus "going steady."

Norms around cohabitation and marriage have also shifted dramatically.[5] Prior to the 1960s, living with a romantic partner outside of marriage was considered scandalous. Now, just over half a century later, nearly seven out of ten adults in the United States believe it's acceptable for an unmarried couple to live together, according to a Pew Research Center study on cohabitation.[6] Today, in fact, the custom of living together *after* getting married is also one that is changing: According to the U.S. Census Bureau, 3 percent of married couples live separately. The increasing ease of telecommunication, the advent of internet dating, and the increasing prevalence of social media in our romantic lives have led to a rise in long-distance relationships that would have been unimaginable sixty years ago.

And, of course, the increasing (and welcome) normalization of once-taboo homosexual relationships has shifted the landscape of love and sex in tectonic ways. Huge demographic shifts make it an exciting time for us to redefine what we want in our relationships and how we understand them. But exciting times can also be very confusing.

We're no longer bonding or building nests together the way we have for the entirety of our species's evolutionary history.[7] Of course, being single doesn't equate to unhappiness any more than being in a relationship guarantees lasting contentment, but this rise in singlehood has collided with another stunning global statistic: At any given moment, nearly one in four people worldwide is experiencing a measurable sense of psychological loneliness.[8]

Public health providers, behavioral scientists, and governments alike have all spotlighted loneliness as contributing to a myriad of alarmingly adverse effects on our mental and physical health. In fact, numerous studies have established a link between our physical health and feelings of loneliness, and some have even shown that social isolation and poor social connections can have negative consequences roughly as harmful as smoking fifteen cigarettes per day.

This loneliness epidemic is a symptom of something that is much bigger and has the potential to trigger unprecedented and stark biological consequences. Our species is on the precipice of what I have come to think of as an *intimacy crisis*.

Intimacy is one of the essential ingredients in our recipe for life. In the absence of romantic and sexual connection, we may survive, but we will not thrive. In one recent survey of over 400 people living in the United States, we asked: "How important is a good intimate relationship to your overall life satisfaction?" On a 10-point scale, roughly two-thirds (67 percent) of participants said it was very or extremely important; only one in fifty said it was not important.

The science agrees: Humans *need* meaningful relationships.[9] Connections. Not simply because they bring us joy, but because we've evolved in a context in which the fabric of our lives is dependent on them.

In the pages to come, we will explore the biological, cultural, psychological, social, and ecological forces that come into play over the life cycle of our intimate relationships — the needs and wants that drive how we make, break, care for, and remake our intimate lives.

We are at the dawn of an age in which technology is fundamentally changing the way people meet and connect. Meanwhile, across the globe, shifts in cultural perceptions are radically transforming our understanding of the spectrum of human intimacy and how it relates to sexuality, gender, relationships, and reproduction. Even in this bewildering modern era, where moments of true human connection are becoming increasingly elusive, the search for intimacy remains the most human of human impulses. We need to feel known, cared for, and tended to. And it is this biological need for intimacy — so much more than sex drive — that illuminates how we date, mate, and ultimately choose to build a life with someone.

INTIMACY IN CAPTIVITY

If loneliness is an epidemic across all age groups and demographics, it is especially troubling for the elderly, who are too often living and dying alone. This reality was brought sharply into focus for many Americans in early 2020, as the rapid spread of COVID-19 forced senior living facilities and care homes across the United States into lockdowns. To protect our vulnerable loved ones, many people had no choice but to visit with aging parents or grandparents through a glass window or computer screen.

As the world came to grips with this "new normal" in the face of a global pandemic, research scientists at the Kinsey Institute instinctively began to think about the impact "social distancing" was also having on our intimate lives. Physical distancing, as social scientists prefer to call it, imposed unprecedented limitations on our romantic and sexual relationships, especially for those who were unable to visit with or were otherwise separated from their partners. At the same time, many of us also wondered what remote work and school closures would mean for those who were homebound with partners or children. Would this time together create new opportunities for couples and families to bond? Or would we break under the strain of this collective crisis?

Rather than just ponder these questions, my colleagues Justin Lehmiller, Amanda Gesselman, Kristen Mark, and I rapidly launched an online study, surveying a diverse sample of thousands of adults from around the globe. Based on an initial snapshot of 1,500 participants, we found that in the early months of the pandemic, nearly half of those in our sample were generally less sexually active, noting that their sex lives had declined during shelter-in-place orders.[10]

On the other hand, one in five also reported having expanded their sexual repertoire by incorporating at least one new activity (sexting, sending nude photos, trying a new position, or sharing fantasies) into their relationships. Some were more likely to try new things than others — the more experimental groups included younger adults, people living alone, those who felt lonely, and those with sensation-seeking tendencies. Another study found that among those with high relationship satisfaction, nearly half reported feeling more connected to their partner after the pandemic began.

All of this suggests that even when faced with drastic changes to our intimate lives, we can find creative ways to adapt and meet the challenges of the moment — a testament to the resiliency of human relationships as we weather unpredictable storms.[11]

We find intimacy where we create emotional connection, experience vulnerability and trust, and engage in mutual care. These characteristics can, and often do, illuminate love relationships, but they are not exclusive to those alone. Sometimes it's a deep and abiding emotional bond that endures long after other relationship ties have been severed, as in the case of a friend of mine who broke up with his boyfriend but continued to have platonic "slumber parties" with him (and his cat) for months after. They slept in the same bed and cuddled while watching movies. There was nothing explicitly sexual or romantic between them after the breakup, but they continued to be drawn to the comfort and safety they experienced within the familiar rhythms of each other's gravitational orbit. Their unconventional friendship satiated a need they both had for intimacy — but it also inadvertently became a barrier to establishing new relationships once they were ready to date again.

Unlike sexual desire — the physiological urgency of which can be dispatched relatively easily via masturbation or sexual activity —

intimacy releases a different, harder-to-come-by flood of hormones and neurotransmitters. Recent neuroimaging studies, such as those by social neuroscientist Bianca Acevedo and colleagues, also bear this out, demonstrating that those in long-term highly satisfied relationships have a specific pattern of brain activation that is above and beyond the patterns we see when assessing the neurobiological underpinnings of human friendships and family relationships.[12]

Studies using functional magnetic resonance imaging (fMRI), a technology that measures brain activity by mapping the flow of oxygenated blood, have found that romantic intimacy involves attachment and bonding systems (such as the globus pallidus, a region of the brain associated with maternal attachment) as well as those associated with romantic love (such as the dopamine-rich ventral tegmental area and dorsal striatum, which play an important role in reward cognition and decision making), and even those associated with sexual motivation and frequency (hypothalamus and posterior hippocampus). Put more simply, when we look inside the brain of someone who is in love, we see a cascade of activity that helps explain why the competing motivations for both social monogamy and sexual variety underwrite so many of the pain points in modern relationships.[13] When we look at the brain, we can see that what we experience as passionate love, an overwhelming force of emotion, is in fact made up of constituent parts. When they are activated together, we feel the warmth of intimacy, of feeling understood and safe. Loved.

But these parts can also be activated independently of one another. Our brains can process good sex, they can process companionship, they can process trust — and we can derive joy from each of these aspects. But they don't always add up to intimacy, or to love.

Why not?

Humans have been attempting to answer that question for hundreds of years. And although we may never fully understand the vagaries of romantic love, there is value in understanding each constituent part, not as a mysterious, esoteric feeling but as a quantifiable scientific reality. From studies like these we can learn that intimacy — that heady cocktail of attachment and sexual attraction, love and lust, trust and desire — is not a fuzzy concept or a metaphor to be approached sidelong but something that can be separated into its components, put under the microscope, and understood as the uniquely powerful force it is.

THE DIGITAL DISCONNECT

In a world of 8 billion people, where so many of us are living in densely populated cities and communities, in an age when a majority of the developed world has access to the internet and mobile technology, it's incumbent upon those of us in the scientific research community to ask, *Why on earth do so many of us feel so lonely and isolated?*

I believe one of the key answers is that the biological mechanisms evolution produced to help our species navigate human connection are flummoxed by the reality of the present moment. Evolutionarily speaking, we are not wired to deal with the deluge of data that defines our modern age. Today we can find endless potential partners on our smartphones, schedule a sexual hookup through an app, use customizable in vitro fertilization to procreate, check in on an ex through social media, marry people who were born thousands of miles away. These possibilities are in so many ways remarkable demonstrations of human ingenuity, but our species didn't evolve in this type of environment, and neither did our sexual or relational behaviors. That's not to say these changes are for the better or for the worse; rather, they are different challenges from the ones faced in *Homo sapiens*'s ancestral environments. Because of that mismatch, the human animal is simply ill-equipped to face today's new interpersonal challenges.

The digital age has opened new possibilities for human connection, but it has also brought new obstacles when it comes to intimacy. People spend more time than ever on screens, and less time pursuing in-person connections. With the emergence of technologies like video chats, short video clips, and the integration of images with sound, movement, and expression on dating app profiles, we are at a tipping point.

Young people are spending more time swiping on apps than meeting potential partners. Data collected over the last decade-plus from our annual Match study have shown unequivocally that more single Americans have met their most recent first date through the internet, particularly on dating apps (and this pattern seems to be occurring globally). By making it easier than ever to search for someone with the interests and the physical traits you most desire, anywhere in the world, these apps allow people from all walks of life to level up their partnering aspirations; after all, the chance of meeting your "perfect match" appears much greater once your choices are no longer limited to those who happen to cross your path. And our expectations have indeed risen. One large-scale recent study of online daters in four U.S. cities (New York, Boston, Chicago, and Seattle) found that both men and women tend to pursue potential partners who are about 25 percent more desirable than themselves.[14] (Desirability rankings were based on the number of messages received and proportion of replies to messages sent, and by whom. If you contacted a much less desirable person, their desirability score would rise; if they contacted you *and you replied,* then your score would fall.)

The results are a mixed bag: Those billions of daily swipes yield an average match rate of less than 2 percent.[15] In the 2021 wave of our national Singles in America survey, Helen Fisher and I found that nearly half of single adults felt technology has made it more difficult to forge real connections; interestingly, Gen Z and younger millennials were the most likely to say this. At least one recent study suggests that those who hold negative views about dating apps and websites carry that negative bias with them into their first dates, where it hovers over their new connections like a dark cloud blocking the sunlight.

New online and app-based dating markets are exploding, especially in Asia and Africa, and in parts of the world where women have only recently gained greater gender equality. Estimates suggest that, globally, more than 350 million people use these dating services every year. And it's no longer just digital natives; older generations are increasingly adopting norms ushered in by tech-savvy young people, in a process social scientists call intergenerational cultural transmission, or the passing of information from one generation to another. Traditionally this happens from the top down; we learn from our parents and grandparents, who learned from their parents and grandparents, and the most useful stories, rituals, and skills are transmitted because they serve a social and cultural purpose.

With regard to contemporary dating technologies, though, this intergenerational pattern has reversed itself, and it has done so with shocking speed given the internet's history as so evolutionarily recent that it's a relative blink of the eye. A journalist from Colorado I spoke with told me her daughter had set her up with a dating profile right before heading off to college, as she wanted to make sure her single mom didn't get lonely in an empty nest. More and more, we are seeing older generations finding their way onto dating sites, but they're not always the ones doing the signing up. Kids take their parents' pictures, teach them texting jargon, and help them navigate unfamiliar online platforms.

For a few years in a row, one of the fastest-growing dating websites in America was OurTime, which specifically caters to those age fifty and over. What's interesting is that people signing up for OurTime are used to different dating norms than younger people today. How do you tell Grandma that the reason her last date isn't texting her back is probably that she's been ghosted? Or explain to Great-Uncle Leroy that he really should keep that private pic a private pic, or instruct him on how many successive texts is too many to send immediately following a date?

These are relatively minor issues of shifting social norms, but beneath the possibility dating apps provide to people of all ages is the reality that whether we are twenty-five or sixty-five, we are losing something when we shift courtship practices to internet platforms. This unprecedented opportunity comes with a cost. In a world that serves up all manner of digital connections, it is easy to forget that we've evolved to eventually need something more tangible.

This need for physical connection is rooted in our evolutionary legacy as social primates.[16] Research has repeatedly shown that physical touch — ranging from an outstretched hand in a business negotiation to a compassionate hug in times of grief — can have enormous impacts on feelings of social connection. And affectionate touch, such as holding, cuddling, and other acts of physical affection, has always been a critical way for us (and our primate ancestors and cousins) to build bonds and alliances, form and maintain relationships, and signal all sorts of behavioral intentions, from care and support to dominance and subordination. That humans are a social animal possessing language as our primary form of communication adds to our complexity as a species, but physical touch is a legacy that is very much hardwired into our physiology.

Intimate touch is key to basic survival for many animals. For most newborn mammals, no longer protected by the warmth and constant food supply in their mother's womb, huddling and cuddling are essential to maintaining body heat (thermoregulation), initiating bodily processes like feeding and urination, and often even for maintaining the body's basic living conditions (homeostasis).[17] Several laboratory studies have shown that baby rats who receive more licking and grooming from their mothers are calmer and respond better to stress later in life. In other words, attachment touch, the kind that forms social and emotional bonds early, leads to better coping later.

In this case, what's true for rats is also true for humans. Affiliative behavior, what we think of as physical affection, has been with humanity from the start. Touch is one of the first senses available to us when we are born. As infants, we experience it almost continuously as we are fed, bathed, changed, and soothed by the adults who care for us. Touch is so critical to our social and emotional development that the absence of touch in our formative years is considered a hallmark of neglect, cited as contributing to an increased risk for significant physical and mental health problems in later life, from anxiety and depression to obesity, heart disease, and multiple types of cancer.[18]

This hunger for close connection and intimacy has been stifled and misdirected in today's digital world. In one national study, we found that over a third of coupled Americans surveyed said they were not touched enough by their romantic partners. Of course, there are also some who feel smothered in some situations, touched too much by a particular person or at a particular time; others are experiencing unwanted and nonconsensual touch. Some try to negotiate their needs and their boundaries, while others languish in despair, feeling unsatiated, unsatisfied, and lonely despite being in a committed relationship. But the fact remains that humans need physical affection, and so many of us feel unsatisfied with how much — or, rather, how little — we are receiving, to the point of startling social, behavioral, and health consequences.

Adolescent depression has been steadily on the rise, with many experts discovering links between increased smartphone use and decreased social interaction.[19] At the same time, frequency of sexual activity, especially among the young, has declined in national studies; this pattern has been referred to as a "sex recession."[20] Some analyses suggest this lowering average may be because of overall reductions in sexual frequency, while others highlight that it may be the result of a greater number of young adults who are sexually inexperienced.[21] Individuals and nations are struggling with their fertility, with reproduction rates down in developed countries around the globe.

All told, it's a grim picture: human intimacy on the edge, with social connectedness of people and populations spiraling downhill, largely because we don't have a collective understanding of who we are and why we need intimacy in the first place. The human need for intimate touch is one of our most intense desires. And when our close relationships are paired with our other evolved romantic and sexual motivations, the resulting emotional and erotic intimacy is a powerful force of nature — even if for some, like my airplane friend Ginny, it can also be a confusing one.

There are three balls we juggle constantly: close relationships, sexual desire, and romantic pair bonds. What our research is only now uncovering is that when we talk about one of these things, we talk about *all* of those things. Understanding how our social worlds are becoming simultaneously wider and shallower in today's technological era will help us unlock the mechanisms of the intimacy crisis — so we can work to overcome it.

Chapter Two

CRAVE

Almost half an hour into dinner, I noticed Elizabeth had barely touched her entrée. As friends and fellow academics, we've known each other long enough that I could tell her mind was elsewhere. Whenever we get together, she and I inevitably end up talking about our romantic lives, and as our conversation wound around to the subject of the guy she had recently started seeing, I noted a shift in her body language.

A wide smile quickly spread across her face. The more she talked about this new man in her life, the more animated, even giggly, she became, almost like someone who's had a couple of drinks. A dopamine rush, the kind we get when thinking of the target of our romantic attention, can make us feel and seem a little buzzed. Her nonverbal behaviors all suggested intense feelings. She was suddenly chatty and effusive. As she spoke, she leaned forward in her chair, eyes wide with focused excitement. Elizabeth was overflowing with joy, almost euphoric, but she wasn't entirely comfortable, either; the pauses in her speech suggested some hesitation.

Interestingly, she refused to use the word *boyfriend* — as if she were allergic to the term. Instead, she maintained they were just "hanging out" and "keeping it casual." I wasn't buying it. In a recent wave of our Singles in America study, we found that roughly one in three single adults who "hang out" eventually end up more involved, a gateway to becoming more sexually and romantically connected.[1]

"It's not that serious! I swear," she insisted, despite having just described in vivid detail how his smile lit up the room, what an incredible dancer he was, and the way his touch sent a jolt of electricity through her body. When she commented that he always smelled "like he just stepped out of the shower after a five-mile hike through a cedar forest," I noticed her cheeks were flushed and her pupils had started to dilate.

"Do you miss him?" I probed. "Like, right this second?"

"Well, yeah, a little. But that's just because I haven't seen him since Sunday."

It was Tuesday. "How often do you talk?"

"Not all that often. Most days? Okay, every day. Well, a few times a day. I don't know — do you count texting?"

As if on cue, her phone, facedown on the table, vibrated. She laughed and shot me a guilty look. "I'm not turning that over," she declared.

I smiled.

"Hey, if *you* met someone whose pancake-making game was this turned up, you'd think about him all the time, too," she added defensively.

"So you stayed for breakfast? Wait... what about that other guy you were seeing?"

"He's... out of the picture."

"Okay. Is your new guy dating anyone else?"

"Well, we haven't exactly had 'the talk,'" she said, the unspoken rest of the thought here clearly being *But he sure as hell better not be.*[2] The urge to "mate guard" — to ward off potential rivals — often begins very early in the courtship process.[3]

"I hate to be the one to tell you this, but it sounds like you're in *love*."

"What? No way," she laughed, waving her hand dismissively. "You think everyone's in love." Her phone buzzed again. "Well," she conceded, "maybe. A little bit."

And then, not surprisingly, she changed the subject; denial staves off the uncertainty of new love and makes us feel we're in control.

Science says we're not.

Biologically, when we unpack romantic love, "catching feelings" can look and feel a lot like a sickness.[4] In the "afflicted," we detect a range of psychological symptoms from focused attention and intense energy to intrusive thoughts, possessiveness, and obsessive thinking.[5] We also see a collection of biological and physiological responses to the object of desire, like sweaty palms, dry mouth, rapid heartbeat, loss of appetite, and being distracted from other tasks. In the brain, we see an increased production of dopamine in the ventral tegmental area, which is the neurotransmitter behind the "pleasure" circuit. Social neuroscientists have even shown that feelings of lust and romantic attraction diminish function in the prefrontal cortex of the brain, which is where rational decision-making takes place.[6]

If you described experiencing these same uncomfortable physical and emotional sensations in a different social context — for example, while taking an exam — we would think you were having a panic attack. As it turns out, scientifically speaking, you basically *are*. When you're under the spell of romantic love, your physiological response is nearly indistinguishable from anxiety.

Of all the afflictions humans suffer, love is one of the most universal. In every human society where anthropologists and sociologists have documented relationships, they have found evidence of romantic love.

While we can't have absolute certainty in the cross-cultural record, there isn't a single known example of a society that doesn't acknowledge experiences of love. And researchers like cultural anthropologist William Jankowiak (who has authored more than a hundred academic and professional publications on love around the world, from urban societies in China and Mongolia to Mormon fundamentalist polygamy) have scoured the record to check.[7] Love's presence is everywhere, in our language, our stories, our artistic expression, and most of all in our behaviors. It is a core human response, what we call a human universal, occurring across time and place. Though there are some notable exceptions (those who identify as asexual or aromantic, for example), the overall pattern shows that people in all societies seek out love, and do so with a force that can feel a lot like hunger.[8]

Like Elizabeth, we feel the push and the pull, the agony and the ecstasy of the pursuit of love — we run from it even as we run right into it.

But what exactly is *it* that we are feeling? Why do we, as a species, crave love?

BRAAAAIIIIINNNSSSS

When prairie voles mate, it's for keeps.[9] They form intense lifelong bonds, they share nest-building responsibilities, and both parents play important roles in raising their pups. Which is why they make excellent subjects for those of us looking to understand the biological forces underpinning our motivation for love and sex.

In fact, research on the brains of the socially monogamous prairie vole, alongside that of the less social, nonmonogamous montane vole, shows how brain structure controls the function of a chemical in the brain that plays a crucial role in pair bonding — the neuropeptide oxytocin. In research pioneered by my Kinsey Institute colleague Sue Carter, what we see when we compare the brains of each species is that the monogamous prairie voles produce more oxytocin and have more and denser receptors in the brain for the oxytocin molecule than the more promiscuous montane vole.[10] Carter's studies were some of the first to look at pair bonding in animals by way of both their behavior and their brains and have led to research on a better understanding of the role of oxytocin in social behavior, especially bonding.

Oxytocin is an evolutionarily ancient hormone that plays a lot of different roles in human biology.[11] It's involved in female reproduction, including in uterine contractions during birth and milk letdown during lactation. It triggers the physiological mechanism by which mothers and their newborns bond during childbirth and breastfeeding. It facilitates social behavior, as in those close bonds between mother and child or between child and another caretaker, which we maintain throughout life. It's also sometimes referred to as the "cuddle hormone" because it's released in the body during affectionate touch and sexual activity, especially during orgasm. It's possible, even probable, that oxytocin is the reason some women experience contractions with intense orgasms, as the hormone activates some of the same systems.

Oxytocin shows up in our neurological response when we interact with the people closest to us. It's critically important in understanding the bonds involved in intimate coupling, but it's just one of the crucial elements in our neurochemical adaptations that allow us to form that closeness — an ancient evolved system, involving a cascade of chemical messengers in the brain (including vasopressin, which modulates behaviors like cooperation, aggression, and risk-taking).

Human beings engage in what's known as preferential sociality, meaning we select particular romantic and sexual partners based on preferred traits. This tendency, along with pair bonding, creates a hierarchy of social relationships, and one must have the neurobiological architecture to support it.[12] Not all species do. It's the evolved architecture of our brain that allows the flexibility to interact with nature *and* nurture in the specific ways that explain human behavior. As behavioral scientists continue to study our neurobiological substrate, which is a way to describe the building-block ingredients in our basic chemical brain soup, what we're trying to do is narrow in on the mechanisms that make us feel safe, calm, even euphoric.

Of course, there's a flip side to that feel-good flush we experience from physical intimacy, and it has to do with the social context. While hugs, massage, cuddling, and other forms of close physical contact can produce a surge of oxytocin that makes us feel good when they are from someone we trust, when they come from an unknown or unwelcome person, the body will generally elicit a fear or flight response. Sometimes wires get crossed in ways that can be confusing and upsetting, such as unwanted touch that is physically pleasurable. If someone experiences climax from tactile stimulation during nonconsensual sexual events, that can cause feelings of guilt and shame.[13] This does not mean a behavior was secretly wanted — it means that the body and mind are not always coordinated in response, and that not all bodily reactions are calibrated adaptations.[14]

As it turns out, oxytocin does a whole lot more than make us feel good. Studies have shown that it acts as a sort of extension of the human immune system, facilitating stress coping behaviors, acting as an anti-inflammatory and antioxidant, and at least one experimental study found that coupled adults with higher oxytocin levels had better wound healing after laboratory-controlled injury.[15]

Love may actually be the best medicine after all.

A TALE AS OLD AS TIME

It's a commonly held belief that humans have only been coupling up for love for the last few hundred years or so. Before that, the thinking goes, humans mostly entered into partnerships for reasons of property, money, diplomacy, or class. We paired up to survive. But evolutionary science tells us a different story. While it's true that societal constraints on pair bonding and the cultural institutions humans build around them have evolved over time (marriage as a negotiated aspect of land transactions, for example), the reality is that love, as we understand it neurobiologically, has been a powerful driver of human-to-human interactions since there have been human-to-human interactions. Our species has always craved intimacy, even if our ancestors didn't realize it at the time.

What has changed, and what is fascinating from the perspective of social and cultural evolution, is the *role* of love in pair bonding over the course of the last 10,000 to 15,000 years.[16]

Let's start at the beginning. Our species's original love story begins about 4.4 million years ago, thousands of generations before anatomically modern humans. Our ancestors were undergoing enormous change as *Homo sapiens* emerged in Africa and then began migrating to all corners of the globe.[17]

Some of our origin story may be familiar to you: The future human lineage had diverged from that of our closest primate cousins 2 million years before, and our ancestors were gradually undergoing anatomical changes. Bipedalism began to emerge, and with it came the social consequences of walking upright on two feet. Think how different communication became, and how much more important facial features and expressions would seem. This changed the ways in which our ancestors could share information and express themselves, and likewise be understood by those around them, deepening the ability to communicate with each other. Sex even changed, as face-to-face coitus was now possible. Forward-facing sex is more intimate: It involves close eye contact, rhythmically mirroring breathing, smelling and tasting each other. It opened the door to sex as a mutual act, one in which physical pleasure is a shared experience. Sex could now become an expression of intimate connection.

Meanwhile, as sociality became more complex and individuals learned to manipulate their environment, our anatomy continued to adapt. Early human heads grew to accommodate our bigger and bigger brains, until our forebearers' newly bipedal pelvises were just about at the limit of our ability to squeeze those heads out during childbirth without causing biological and medical problems. This is one reason that, compared to other primates who give birth unassisted and with relatively little difficulty, humans struggle during childbirth to the point where we need other humans to assist to ensure that both mother and child survive the process. What's more, our large-brained infants are essentially helpless, dependent totally on parents or caregivers, for the first two-plus years of life. In terms of security, this would have been (and still is) an extraordinarily difficult task for single members of the species to regularly undertake alone.

Anthropologist Sarah Blaffer Hrdy, whose work highlights the unique role of women and mothers in the evolutionary process as they compete and cooperate to shape survival and reproduction for themselves and their families — a powerful dynamic we'll revisit in the chapters to come — has suggested these biological factors are why we are cooperative breeders, meaning we rely on the help of experienced "alloparents" (a network of family and friends who provide parental care) to ensure our young survive and thrive.[18]

Since we were unlikely to escape the need to have other people around to facilitate childbirth and rearing from the very beginning, the process of evolution selected a way that maximized our chances at reproductive success.

Enter pair bonding.

Evolutionary scientists believe pair bonding evolved among our ancestors right around the same time as bipedalism.[19] Coupling up for cooperative reproductive purposes would have fundamentally changed mating, allowing for mutual acquisition of resources including hunting and gathering, and eventually dual-parent care of infants, protection of females during pregnancies, and shortening of intervals between births, as one parent no longer had to wait until a first baby was more or less self-sufficient to care for another. As our ancestors conquered their new environments, moving from the trees into the savannah, the adaptive strategies that maximized their reproductive success involved *families* — overlapping, cooperating groups with a two-person dyad at the center.

Many species design their own environments. Look closer at that anthill in a crack of the sidewalk, or better yet, those birds building a nest in your backyard (the birds are probably pair-bonding, as almost all birds do, at least for a breeding season). But compare that kind of small-scale change to the grandeur of the Hong Kong cityscape or the Panama Canal. This remarkable ability of humans to manipulate and master our environments on such a spectacular scale is a direct result of cognitive evolution, which is the result of our large brains, which is the result of women being able to birth our large heads, which is the result of complex cooperative sociality, which is really the result of pair bonding.

In other words, the evolution of pair bonding—romantic attachment, as we think of it today—set humans on a path of anatomy and sociality that allowed us to become, for better or for worse, the most dominant and invasive species on the planet.

It's worth reminding ourselves that evolution is not teleological, meaning we are not evolving "toward" something. Nor did we evolve to be a "better" ape or chimpanzee—chimps evolved for their environments, and our ancestors for theirs. Rather, a species divides and survives because it's doing what is most adaptive, what leads to more survival and reproduction relative to others in a specific ecology.

And what has been most adaptive for *Homo sapiens* over many millennia is social monogamy. We evolved to form strong pair bonds with a single mate because this two-parent unit, working together to raise, protect, and care for their offspring, led to stronger, smarter, and healthier young.

Crucially, however, *social* monogamy is not always in lockstep with *sexual* monogamy.[20] And there's the rub. So much of the misery we have felt in our romantic relationships as a species—so much of the psychological torment—derives from this confounding paradox. Our unhappiness in love often points back to our attempts to make (or keep) a socially monogamous relationship sexually monogamous, or vice versa.

The evolution of human mating, encompassing selection for mating systems and reproductive strategies over millions of years, has given us a clear answer about what works for the species, and that answer is social (but not necessarily sexual) monogamy.

THE EVOLUTION OF COURTSHIP

After years studying the evolution of human biology and behavior and examining the social and sexual patterns of societies around the world, I've come to believe that the two greatest changes to human courtship in the last 4 million years—that is, since the evolution of social monogamy—are the agricultural revolution and the proliferation of the internet.

The impact of these two changes during the inexorable march of civilization is marked by the way they changed the human animal's ability to interact and the market forces at play in our social lives — governing how we allocate everything from food resources to territorial borders to ease of social and material opportunities. This in turn altered patterns of expressing our evolved need for intimacy. These two monumental events in human history didn't change the desire for love and sex, or the reasons we engage in courtship and marriage, but they did change *how* people came to practice these things.

Let's start with our ancestors, going back five hundred generations and piecing together their reaction to the first agricultural revolution, which is sometimes called the Neolithic revolution or Neolithic demographic transition. About 12,000 years ago, some human populations began transitioning from hunter-gatherer to agriculture- and settlement-based societies.[21] Ultimately, this allowed for the domestication of plants and animals, which meant that families, now able to produce and store excess food, no longer had to wander in search of sustenance; instead, they could settle in one place, which led to the development of permanent structures regarding knowledge, trade, and government. In other words, greater stability. But this shift to more permanent settlement and farming resulted in new forms of inheritable wealth and competition, leading to unequal access to these new resources. Land, livestock, and currency became inheritable capital, creating a new avenue for investing in offspring and one's evolutionary legacy — in addition to food and safety, investing in, acquiring, and inheriting capital became interwoven with courtship and marital arrangements. In turn, human social organizations (and marital agreements) became more complex, offering more opportunity for resources beyond the immediate family and kinship network.

As people became less mobile and more sedentary, social life changed, as did the economic factors involved in mate choice. Suddenly land and territory played a larger role in human courtship decisions. The work of my gender studies colleague Colin Johnson reminds us that in rural parts of America, for instance, dating and mating were sometimes thought of like animal husbandry, in that you chose partners based on their pedigree and them being "good stock" to match with, even while there were indeed things like casual sex happening out of public sight.[22]*

It's a popular view that human sexual behavior was indiscriminate prior to the agricultural revolution, and that it was this new social organization that allowed for the introduction of land ownership, which shackled people to relationships and in turn enabled social monogamy to become the norm for the first time in history. It's a compelling argument, but there is very little scientific evidence to support it. The truth is, people were socially monogamous long before land ownership.[23] There is genetic and archaeological evidence that early humans formed socially monogamous pair bonds not unlike the ones we form today as early as 4 million years ago, when they were still largely hunter-gatherers. The agricultural revolution, 12,000 years ago, probably had a more nuanced effect on relationships by institutionalizing pair bonds into regulated marriages tied to resources, a process that exacerbated gender inequality in heterosexual contexts and also probably took some of the passion out of courtship and romantic relationships. But these economic developments didn't make us monogamous — they merely reinforced what was already happening.

The evolutionary legacy of human pair bonding has long been influenced by a combination of both individual preferences and the preferences of our kin, with the relative influence of each waxing and waning across our ancestral past. In the end, most people the world over have had a partner to share some aspects of their lives with, generally including sexual and reproductive matters. People are not always paired by choice or because of attraction, but they form intimate bonds nonetheless.

---

* Human history has demonstrated this most startlingly in the context of same-gender passions, which have so often been forbidden and forced underground, or when permitted often relegated to a secondary status behind heterosexual partnering. Even among those civilizations, such as the ancient Greeks circa the fourth century BCE, that we think of as being more sexually permissive, historians and classicists have argued that heteronormative expectations nevertheless prevailed.

We see evidence of this in arranged marriages, which many societies have practiced, in some version or another, for millennia. In one study of nearly 200 hunter-gatherer societies, 85 percent practiced some form of marital arrangement initiated by parents or other close kin. Moreover, the majority (80 percent) of the sample also practiced exchange of resources between marrying families, sometimes referred to as a bride price or dowry. Another analysis, using techniques from population genetics, suggests that arranged marriages have taken place at least since the initial migration out of Africa by modern humans.[24]

Love has always been present in our biology — an inescapable part of our neurochemistry — but for many of those in arranged marriages, love and intimacy came *after* partnering up.[25] In other words, we are the descendants of those who may not have always been partnering for love, but they often, especially if they stayed together, *became* loving in their partnerships over time. Of course, this continues in various forms to this day — from well-meaning friends setting up a blind date to professional matchmakers or a formal arrangement between families.

Romantic love has always been a part of the human condition; what's changed is the order in which we expect things to happen. Now we expect this sequence:

1) Meet a potential partner.
2) Fall in love.
3) Marry or otherwise commit to a monogamous partnership.

But for millions and millions of people in somewhat recent history — and, to be clear, we're generally talking about the straight/hetero cisgender-normative paradigm under which much of the research has been done — the marriage contract was the more immediate goal, with #3 coming before #2. What's interesting is that it doesn't seem as though the order of operations has affected the sum total of love or intimacy that exists in human pair-bond relationships.

In at least one analysis, after several years of marriage those couples with an arranged marriage reported nearly the same levels of satisfaction as those who chose their own partners through courtship.[26] But there is an important caveat to these data: Arranged marriages that thrive seem to work because they occur in cultures that readily endorse them, which at least in part explains why the arranged marriages attempted by contemporary Western couples on today's reality TV shows so often appear to not last long at all.

I see marriage, and even a formal engagement proclamation, as a way to institutionalize our commitments. This includes the resources that often accompany marriage proposals and contracts — whether that's in the form of a diamond engagement ring, a monetary exchange, a parcel of land, or three goats. In other words, marriage is a social contract that structures partnering.[27] But pair bonding and love are not so easily regulated. The way our brain, body, and mind come together to form the feelings we have about a person we want to be partnered with is a force that often transcends the documents we sign to formalize these relationships.

Pair bonding is an evolutionary adaptation. Marriage is a social convention. The two are distinct but not mutually exclusive. In the absence of rules and structures, intimacy remains — our species would pair-bond even without the convention of marriage. And we did so, for millions of years. Marriage just provides a new arena, with rules, for intimacy to play out in; it can grow and develop there, it can wither and die there, and — as in many arranged marriages — it can even be born there.

Marital arrangements try to regulate two distinct aspects of our intimate lives: whom we partner with (and, in turn, build a mutual life with) and whom we engage in sexual activity with (and expect reproduction with). Sometimes sexual monogamy is an expression of intimate love: we are sexually fulfilled by our pair-bond relationship and don't want to be sexually involved with others. But other times it is the marital agreement, not necessarily our feelings of romantic love, that attempts to define what is acceptable, and in our culture that agreement generally carries the expectation of monogamy in all its forms.

Which brings us back to one of the most essential findings from my entire research career so far: It is the tension between our biological motivation for love and intimacy and our simultaneous motivation for sexual pleasure and exploration that vibrates at the heart of all romantic relationships.

LOVE IN THE AGE OF ALGORITHMS

The second big shift in the evolution of human courtship occurred as recently as the mid-1990s, with the launch of the first modern dating websites. By the early 2010s, dating apps were well on their way to becoming firmly established in the mainstream. Then in my twenties, I had already become versed in the courtship rituals of these apps, but for my cousin Jen and her husband, Dave, they were entirely novel. They had fallen in love a decade earlier, before the rise of internet dating, and found each other the old-fashioned way, through mutual friends and shared hobbies. On more than one occasion while I was hanging out at their place, if Jen got hold of my phone, she and Dave would check out my profile and marvel at all the potential dating options available with just the swipe of a finger.

Different generations have different dating experiences. Dave's parents, who are Japanese, are a wonderful example of this. Before Dave was born, his father was married and had two daughters, but his first wife died of breast cancer. His family encouraged a second marriage, in part because there were small children involved, and arranged for a new wife, who would eventually become Dave's mother, to move in with Dave's father in America. They were married for over fifty years and built a lovely life, and she cared for Dave's father as he developed dementia and ultimately died at age 101. Love at first sight isn't required, or even typical, in order to have a satisfying and loving relationship.

But what does the pursuit of that constellation of feelings we call "love" look like when we stop relying on all five senses and our cultural context to choose a mate? I've heard hundreds of stories about potential matches that fall apart once the senses of smell, taste, touch, and so on are invoked. I interviewed one woman who confided she still wondered about her high school boyfriend, who had all the qualities she wanted in a mate: looks, intelligence, and a great sense of humor. He was crazy about her and went on to become an MD with a thriving medical practice. She had wanted to be with him, but there was something about the way he smelled. Not that he smelled bad, but she just couldn't get past it. "Pheromones?" she asked me, referring to the chemicals we produce that function like hormones outside the body, with cues to our genetic makeup and immune system function, that can be detected by others of the same species. The answer is a bit complicated, as the question of how well human olfactory systems can detect pheromones specifically versus other factors has been debated — but the short answer is yes, a person's smell, even if not technically pheromones, matters considerably.[28]

People react strongly to an affront to their senses. Two friends of mine, for example, seemed beautifully matched in terms of intelligence, empathy, attractiveness, interests, and goals. After flirting for months, they finally had their first kiss. It did not spark fireworks. In fact, one of them shared with me that the taste of cigarettes on the other was such a turnoff that the drop in his interest, both sexual and romantic, was so precipitous he could practically hear it hit the ground.

This isn't uncommon. A 2007 study showed that roughly half of college students said they have kissed someone and known immediately that there was no chemistry.[29] Many researchers have speculated that this is because we transmit chemical information in our saliva when kissing, but there's very little evidence for this argument.[30] What kissing can tell us is whether a new partner is compatible — if they pick up on our comfort, tempo, and interest, and know how to adjust accordingly.

In addition to engaging our senses when assessing potential partners, we also cannot discount the fact that over the course of human history we have always had at least some indicator of the community they might belong to, which is an equally important signifier. In the past, we dated within our communities, and although proximity might have limited our options, it also provided us with knowledge and shared networks. If you lived in a small town, chances are you knew the person you were dating — or you knew people who knew them. You had a shared understanding of the cultural context.

Online, our community is not defined by proximity — or if it is, that community is too wide-ranging to provide the knowledge we might once have gleaned from mutual acquaintances and family history. Online, we usually date outside of our networks; meeting people we would not otherwise encounter is, in many cases, the whole point. And the geographic range of dating apps further decreases the likelihood of shared community. Even if you limit your search to people within twenty miles, there is no guarantee that a potential match is in the same town as you, and there is no guarantee that they truly live there or, if they do, that they have lived there long.

A colleague of mine confronted this very issue when a potential date he'd met on an app suggested they meet at a local bar in a small town about thirty miles from where we live in Bloomington. Although he was interested, as an African American man, he'd encountered racism in some rural areas in the past and was wary of finding himself in a similar situation. He decided against meeting at the bar because it wasn't a community he felt confident he could navigate safely. Evolutionarily, having no knowledge of the person we're about to encounter increases the risk, just as it felt like it might for my friend in the current political climate.

The most common story I hear about the search for love in the digital age is that people who meet on apps before meeting in person find their dates to be visually different: taller, shorter, fatter, thinner, lighter, hairier. There are biological reasons that people tend to misrepresent themselves. First, the data show it's not that people generally lie about their looks; it's simply that we try to put our best foot forward. Contrary to what you might think, dating app users don't tend to lie about easily measurable things like their height; it's too easy to be caught. Instead, they tend to accentuate the positive. Just as you wouldn't post a picture of yourself doing a keg stand on your LinkedIn when looking for a job, people on the digital mating market are loading photos that present them in the most attractive way, where the lighting is soft and the angle is flattering.

This is understandable, given that the visual region of the brain is closely tied to our mating system. We make a lot of instantaneous decisions about the things we see, and attraction is very much tied to the visual cortex. Dating apps have capitalized on this — it's why we swipe on the person's picture, not their list of interests. It's also why many dating apps are actively developing broader technologies to help users present themselves in the most favorable light.

A person's looks are usually what pique our initial interest. But successful daters don't swipe based on looks alone. Even more helpful than photo filters to make you look younger or thinner, I would argue, has been the rise of the virtual dating coach: services that (for a fee) can help a user "optimize" their profile to include more relevant and enticing data for potential partners. Some apps have also experimented with allowing friends to weigh in on your choices, or limiting how many connections you can make per day in order to force users to slow down, focus, and prioritize thoughtful partner choice.

That said, the reason that, despite their sophisticated algorithms, the dating apps don't always deliver what people initially hope for is that they are simply an insufficient platform for the human nervous system. In the past few years, we've seen more dating apps work to address this gap by trying to incorporate more sensory information for our brains to process, including short clips and recordings of a potential partner's voice, and even encouraging video dates on the platform. Still, given that it is physically impossible to smell or touch another person virtually, the most sophisticated apps can only offer a few of the senses for us to evaluate, meaning they provide less "data" than we would take in from real-world interactions. Really, these matching technologies are in competition with the incredible human brain, which is the best possible computer for processing attraction and connection; it evolved over millions of years to synthesize endless stimuli and let us know what we want, whether we comprehend it intellectually or not. But we can still learn to use the promise of dating technologies more efficiently and effectively.

One paradox of dating apps is that despite their limitations, research suggests that over the long term they still yield desired results.[31] Though it may take time for users to find desirable matches, studies have shown that those couples who met on dating websites and decide to keep dating enjoy similar stability in their relationships to those who met in person. And a more recent study found that those who met their partners through a dating app, compared to in-person methods, were just as strong in terms of reported relationship quality. Yet many struggle to harness the promise of these dating technologies.

Perhaps the single most powerful thing we can do to manage our own experience and the amount of data we receive when we first meet someone on a dating app is to slow down. Allow your brain to process the information it has been provided — read the profile carefully, look at the pictures critically, take the time to evaluate the glimpses of personality the person on the other side of the screen has included (wittingly or not). Are they funny? Shy? Do they look like they're having a good time in their photographs? Do they seem to have interests similar to yours? So much of the online dating experience is oriented toward efficiency. But that means that we often don't spend enough time interacting with a potential match's profile, allowing our already highly efficient brains to actually process the massive amounts of data available to us.

Online dating is efficient, but like intimacy itself, it takes thoughtful effort. And the effort we put in impacts the outcomes: Research has shown no difference in relationship quality between couples who met online and those who met in person. Even in a decidedly modern dating world mediated by technology, it is the old-fashioned, analog dedication to building our relationships that actually dictates whether or not they succeed. When we understand that there is no magic wand to wave to be connected to our "perfect match," no secret love potion for endless satisfaction, we can allow ourselves to enjoy the uncertainty and the journey of unpacking and fulfilling our desires. And this holds true for one of our diverse cravings that engulfs intimate decision-making and relationship trajectories more than almost any other: sex.

ON PLEASURE AND PAIR BONDS

The World Health Organization considers sexual pleasure and orgasm part of sexual health, yet too often sex is left out of discussions of our physical and mental health.[32] According to one study, less than half of gynecologists ask their patients about their sex lives. In Western societies, we often don't talk about sex when it matters: to our physicians or even to our partners. Many sex educators have been trying to change that, and so the conversation has often turned toward how we might make sex more pleasurable. More satisfying. More consensual. More balanced.

Men tend to orgasm at higher rates than women, and up to a quarter of women are anorgasmic, meaning they never or rarely orgasm (the debate in the field of sex research on the causes of the gendered orgasm gap is a hot one).[33] This is all wonderful for a modern effort to understand sexual diversity, but it's not necessarily reflective of the way sex has been thought of over time or cross-culturally. In many parts of the world, sex has a primarily reproductive function. That is, sex isn't always conceptualized as an act of pleasure.

Pleasure is only a sliver of the possible purposes of engaging in sexual activity with a partner.[34] Take, for example, research with Aka foragers and Ngandu farmers of central Africa, where the Aka average a sexual frequency of three times per night with two days of rest, and the Ngandu average two times per night with three days of rest afterward. Among the Aka and Ngandu, sex is work, specifically intended to meet the cultural expectation that children are of primary importance. While it may be more pleasurable than their daytime work of foraging or farming, sex is not conceptualized among these tribes as being central to relationship function or stability.[35]

Renowned evolutionary psychologist David Buss and clinical sexologist Cindy Meston published a major study that attempted to answer the question "Why do people have sex?" and came up with 237 reasons, most of which implicitly assume sex is happening within an ongoing romantic pair bond.[36] The answers ran the gamut, from "I wanted to experience physical pleasure" or "I wanted to get closer to God" to "I wanted the person to feel good about himself/herself" and "I wanted to get back at my partner for having cheated on me." Unsurprisingly, the number one reason for both men and women was "I was attracted to the person."

This is not necessarily a shallow reason for sex! Nor does it mean that our connection with that person will necessarily be "just about the sex." When our evolved preferences are triggered and we find ourselves attracted to someone, both our brain and body often want more: to spend more time with that person and connect with them, both emotionally and physically. Though a sexual experience is not always a shared bonding experience, it certainly can be.

We need intimate relationships to survive, but we don't settle for just anything. And sometimes our wants are in conflict, as when we want social and sexual monogamy from our partners but do not wish to abide by those same restrictions ourselves. And sometimes we fall in love faster than we care to admit. For example, my friend Elizabeth denied she was falling in love with her new guy because she was scared of getting hurt; she was trying to pull the kill switch on her own ride. Luckily for the species, that rarely works. Psychological research has shown that people vary in how easily and quickly they fall in love, a reminder that we all operate on our own intimate timeline. But when love does arrive, it is a neurobiological response; denying it is like insisting you're not hungry when you haven't eaten all day. Elizabeth and her new beau "hung out" for almost another six months before they moved in together. I recently got an invitation to their wedding.

All of this is to say, there are a variety of intersecting reasons why we might let our guard down enough to be emotionally intimate with a sexual partner — or why we might want to keep them at an emotional arm's length. Of course, there's more to human intimacy than this binary choice. What we need from our intimate relationships is complex, predicated on a whole host of evolved biological and social needs, but what makes the intimate animal truly unique is our cognitive capacity to modify how we fulfill those needs based on what we *want*.

Broadly speaking, it has long been assumed that while we "need" sex, we merely "want" intimacy. But I would argue that in fact we want and need both, and that *both* sex and the full cocktail of emotional and physical intimacy are crucial not only to our individual happiness but also to the survival of our species. What we want — what we *crave* — is part and parcel of what we need.

Identifying the nature of our desires is step one in understanding our relationships and achieving our intimacy goals. As we move into the next phase in the relationship life cycle, the challenge becomes figuring out how we can wield this knowledge in the real world when we begin in earnest to search for a mate.

Chapter Three

SEARCH

Once a year, in the shallow, temperate waters off the coast of southern Australia, male and female weedy sea dragons seek each other out in the coruscating light of the early summer sun. Spinning their translucent fins and using their tails as rudders, individuals pair off and begin a graceful dance through gently swaying beds of sea grass, towering kelp forests, and rocky reefs. Swimming side by side for hours, they mirror each other's movements in perfect synchronicity. This magical pas de deux ends as a pair spirals upward and the female deposits her eggs on the underside of the male's tail — in the same family as seahorses, weedy sea dragons are one of the few species in which the males give birth.[1]

I have always been particularly fascinated by the elegant "mirror dance" of the weedy sea dragons. So much so that a sea dragon figurine gifted to me by a close friend in graduate school has traveled with me from New York to Indiana and still sits proudly on a shelf in my office at the Kinsey Institute.

The internet abounds with videos of these kinds of spectacular courtship displays across the animal kingdom. With a few keystrokes, you can watch majestic Japanese red-crowned cranes leaping as they raise their long necks to the sky, then flap their wings and bow toward each other. Or there's the "rumble rumps" of the male peacock spider, who attempts to mesmerize the deadly female by elaborately shaking its vibrantly colored tail fan — a movement that, in at least one video meme, is cleverly cut to samba music. The greater sage grouse exhibits a mating behavior that my undergraduate students tell me is remarkably reminiscent of behavior observable on a Friday night at a popular sports bar in our college town. These large, chicken-like birds will congregate in groups called leks where the males display their evolutionary fitness and compete for the attention of the females by strutting, brandishing their spiky tails, and inflating the yellow air sacs on their chests to make a loud *wup* sound. If sage grouse lekking involves a lot of *wupping*, frat boy lekking involves a lot of warm beer.

These courtship displays can be charming to observe — or bizarre, as the case may be — but, more importantly, we can use them to draw inferences about why certain behaviors have evolved to look the way they do and then extrapolate those ideas to explain the interesting ways our own species seeks out partners.[2] And dance, it turns out, is a terrifically common behavior when it comes to attraction, even among humans.

Ginny, my airplane buddy from the flight to Vancouver, described experiencing a moment of surprisingly deep connection with a total stranger that reminded me of the intimate dance of the weedy sea dragons. She had been invited by some friends from her dance studio to a dance festival in the Caribbean, so she took time off from work to spend several days bouncing around a full schedule of workshops and performances. "It felt so good," she told me, "to spend a week focusing entirely on the physicality of my body with other dancers."

On the last night of their trip, Ginny and three of the women from her group wandered into an open-air club on the beach. She felt the booming bass and pulse of the music in her chest from fifty yards away, her heartbeat shifting to match it. They made a beeline for the dance floor. Almost immediately, she locked eyes with a dark-haired man on the other side of the crowd and found herself moving in his direction.

Ginny had no intention of ever speaking to this man after that night. "I'm not ready to give up on my marriage, but in that moment all I wanted was to dance."

So she did. They did. All her senses were engaged as their bodies moved together, her arm suddenly around his neck. She felt his hands on her hips. Tasted his sweat on her skin. The sound of his voice in her ear between songs made her heart beat faster. He smelled like freesia. By the time the lights came up, Ginny was startled by the intensity of her attraction. This man, whose movements were so in sync with hers, seemed to know her so well — except that they didn't know each other at all.

DANCE DANCE EVOLUTION

Whether we're talking tango or two-step, dance can be a key component in developing attraction. Many of us have experienced firsthand that it can trigger a tidal wave of sexual desire — as it did for Ginny.[3] But we can also learn a lot about each other when we dance. By watching the way a person moves and keeps a beat to music, we can sense whether their rhythm aligns with our own and assess their physical compatibility — in some cases, even their personality and general health. While dancing with a partner, as we touch and hold each other, inhaling the other's (potentially) pheromone-packed sweat, we match each other's heart rate and movements in a way that often speaks louder than words.

That feeling of connectedness on the dance floor is what scientists call behavioral synchronization, which is very much a part of how evolutionary selection pressures have shaped social behavior.[4] In the natural world, we can see this adaptation executed to perfection in the stunning displays of a starling murmuration, in which a flock of thousands of birds will swoop and dive through the sky with such magnificent precision they appear to be linked by a collective consciousness that allows them to move through space as a single being.

In humans, when behavioral synchronization goes well, it leads to smooth social interactions, even an almost trance-like feeling of connectedness. When it goes badly, however, a lack of synchronization can make things awkward. Think about a time you were on a conference call where, unable to detect the full range of other people's nonverbal gestures, you became out of sync and found yourself unintentionally but repeatedly talking over someone else. It's even more uncomfortable when this lack of coordination happens in person. Maybe you've experienced an awkward silence during a job interview or the discomfort of being in bed with a lover with whom you can't quite hit the same rhythm. That disconnect is not unlike being on the dance floor with someone who appears to be hearing entirely different music. In the absence of synchronization, a social interaction can go from pleasant to uncomfortable, and even painful, in a snap.

What Ginny was experiencing at that island club, where she danced her way over to a man she was drawn to from among the throng, was like a double shot of behavioral synchronization. "It was like my body knew him," she reflected, "even though we had never met before. It was so much more intense than just a flirtation. We had a connection."

She wasn't wrong. Biologically, our movements are made up of rhythmic patterns and pulsating responses in our physiology that cause our neurons to fire and our muscles to contract and expand. This is partly why it feels so powerful when you connect with someone on the dance floor. As you start to sweat and look into each other's eyes, bodies pressed close, you are syncing at an almost cellular level that allows you to mirror each other's movements and connect. You are in the same moment, in the same song, and in the same space.

It's an intimate experience.

Ultimately, what draws one human being to another — whether that plays out across a crowded dance floor, through a dating app profile, or on a blind date with your grandmother's neighbor's nephew — is a feeling of connectedness that is not unlike the behavioral synchronization that bonds a pair of weedy sea dragons to each other. When you "vibe" with someone, as my students say, it's because you are experiencing a sense of connection on a physiological level.[5] Our autonomic nervous system, a network of nerves that regulates a number of bodily processes — from sexual arousal, heart rate, and blood pressure to respiration and even digestion — has evolved to respond to other human beings in both positive and negative ways. So, just as you might vibe with one person, another might give off "bad energy."

Sometimes it takes a certain combination of people to sync, finding the right yin to your yang. But research has also identified a subset of people who are "super synchronizers," those with an uncanny ability to identify and respond to the behaviors and bodies of others, synchronizing to the physiology of whoever is in front of them. Foreseeably, people with this unique and powerful social skill were rated as more attractive in a speed-dating experiment.[6]

Interestingly, this flirtation can unfold without explicit spoken words. When your nervous system's attention is homed in on one person, even momentarily, that focus and effort to sync to another can be mesmerizing. Likewise, when it's clear that another person is attempting to sync with you — or, even more thrilling, succeeding at it — their efforts can feel like an intimate proposal to become one. To both want and be wanted by the same person can be intoxicating. The American poet Robert Frost was on to something nearly a century ago when he wrote, "Love is an irresistible desire to be irresistibly desired."[7]

This physiological connection is an important component in the experience of desire, an experience that is inextricably bound up with our drives for both intimacy and sex.[8] Desire tangles these drives, sometimes pleasurably, sometimes positively — but also riskily. It gives rise to the pair bonds that give our lives meaning, but it also can threaten the pair bond we may already possess — or lead us into a pair bond we might not choose with a clearer head. That is its danger, and also its reward.

"I didn't go there looking to hook up with anyone," Ginny had insisted. "I never thought of myself as the type of person to have an affair — until we danced."

In fact, she and her dance partner hadn't gotten sexual that night. When the lights came up, they said goodnight and Ginny went back to her hotel alone, though the lingering attraction she felt, still palpable when she shared her story with me, was such that she still wondered what would have happened if she had given in to her desire — just for one night.

THE RISE OF HOOKUP CULTURE

From "hanging out" to "hooking up" to being in a "situationship," modern courtship is much more an exercise in choosing your own adventure than the traditional American date-marry-have-a-baby paradigm of old. My lab has studied hookup culture for years, and we have seen how what was once a transgressive way to dip a toe into the search for intimacy came to be just one more available avenue for sex, and in some cases love. While there are a variety of terms for and subtleties to different types of casual sex, we can define hookups as sexual encounters that take place without any promise of, or apparent plan for, a more traditional romantic relationship — encounters that may include kissing, oral sex, and/or penetrative intercourse, though not necessarily in that order.[9]

I first got interested in hookup culture — professionally — in graduate school. I couldn't help but notice that it seemed like uncommitted sexual encounters were becoming more and more common among college students — who were more openly talking about their one-night stands, too.[10] There wasn't much scientific literature available on the subject back then; psychologists and sociologists had looked at short-term and long-term relationships, but most of the information on hookup behavior was either purely descriptive or came from a public health perspective largely considering casual sex and HIV prevention in populations of men who have sex with other men.[11]

My first empirical study in graduate school on the topic of uncommitted sex examined college students' multiple motivations for engaging in sexual hookup behavior.[12] The lines of research on how we construct our sexual identities are fascinating and complex, so, for me, peeking under the hood of hookup culture in a university environment was a natural progression. (A few years later, I would tap colleagues in developmental psychology, anthropology, and gender studies to conduct the first major review of the academic literature surrounding hookup culture in an effort to further answer the questions that started me down that road.)[13] At the time, hookups were becoming more ingrained in popular culture. We had rom-coms like *No Strings Attached* and *Friends with Benefits,* both movies about friends who become involved sexually but try not to become involved romantically, and LGBTQ+ television shows with same-gender representations of uncommitted sexual behavior, like *The L-Word* and *Queer as Folk.*

So we were not shocked when our analyses found that casual sexual encounters were becoming more and more normative among young adults in North America. Data from the decade following the turn of the millennium suggested that between 60 and 80 percent of college students had engaged in some kind of hookup behavior at least once. And almost all of them knew about, talked about, or at least occasionally thought about sexual hookups, signifying a marked shift in traditional attitudes toward uncommitted sex. If hooking up wasn't actually more common than in past generations, I thought, it was at least far more visible and culturally represented in music, movies, articles, and other indicators of social mores.

Of course, millennials and Gen Z didn't invent the hookup, but a key difference between today's culture and the "free love" of the 1960s and 1970s has to do with record-keeping—the older phenomenon simply wasn't as well investigated by quantitative researchers, so we don't have great longitudinal data. What we do know is that during that time, on average, people were still getting married, and doing so much earlier than today's youths, which makes it partly a demography question. It could be that because young people today are spending greater proportions of their lives single, this simply gives them more time to have more kinds of casual sex. There's also an important distinction to be made between casual sex behaviors and hookup culture. Today, young people talk relatively openly about sexual events that occur outside of committed relationships. It's not done in secret and doesn't involve shame in the same ways it did in the past (under the purposely laden term *premarital sex*). Behaviorally, this all looks more like a cultural sea change than a different perspective on the same phenomenon.

Headlines have asked whether casual sex among youth is a crisis or a symptom of a dying sense of romance. Consider, for example, the excellent book by sociologist Lisa Wade, *American Hookup: The New Culture of Sex on Campus*.[14] Wade argued that hookups are based in feigned indifference — that betraying feelings or need for connection might make young adults seem "clingy," and so her participants report distancing themselves from each other, often precipitously. This is an interesting and important interpretation of the data. Puzzlingly, this doesn't sound like an animal with a deep biological need for intimacy.

Over the last few years, the literature on hookups has proliferated, which has made it easier to comprehend what might be going on in our brains when we seek out "no-strings-attached" sexual encounters, despite our ancient biological craving to form deep connections. So the questions then very quickly become: If humans are biologically hardwired to form intense pair bonds, as we know we are, then why are young people having all this uncommitted sex? How are we negotiating our desires for intimacy and committed romantic love in a landscape characterized by casual sex and casual dating? Are we in danger of becoming a post-intimate animal?

To search for answers, my team and I surveyed young adults who had engaged in hookup behavior about their choices. One of our findings was that the majority of men and women felt, well, pretty good about it later, belying the traditional notion that regret is a mainstay of casual sex. More to the point, however, we also found that although just 6.5 percent of participants went into a hookup situation *expecting* it to turn into a romantic relationship, over a third of them indicated they *hoped* it would.

The inclination for pair bonding is not disappearing; it's just changing shape. In other words, romance may not be dead, and there might be more strings attached to hookup behavior than we initially thought. We may in fact be seeking intimacy through casual hookups — wittingly or not. The proof is in the data: In another of our studies, one in three people have had a "casual" sex relationship become something deeper and more lasting.

Intimacy is not a zero-sum game. Over the course of our lives, we seek it in a number of different ways, not all of them necessarily sexual. We might find intimacy through hookups, through lovers, or through established, long-term relationships. But we also might find it through deep friendships, through family bonds, or even with a beloved pet. By exploring the full spectrum of intimate experiences, we can gain deeper insights into why we want what we want, and how and when our wants vary when we search for a mate — whether it's forever or just for one night — and perhaps even understand to what extent biology determines whose mating stock soars and whose flatlines.

GAMING THE SYSTEM

Rapidly proliferating technologies and lonely modern hearts sometimes make for strange bedfellows. Take, for example, a single-and-dating colleague with whom I shared a flight back to Indiana after a conference. We got to the airport in Indianapolis, about forty-five miles north of Bloomington, and he asked me to wait a few minutes so he could open his dating app. Who am I to begrudge a colleague a shot at love? I was happy to wait. Especially because I understood what he was doing. It was a clever strategy.

The algorithms of many dating apps are based on how many miles you are from a potential match. While you can set the geographical distance for your search, apps like Match, Hinge, Bumble, Tinder, Grindr, and many others inherently privilege people in cities who don't need a hundred-mile radius, as you might if you live in a smaller town like Bloomington. In a major city, you could potentially restrict your search to your local area, like the New Yorkers I know who set their radius to two or five miles, because the density of the city presumably offers ample options. Opening the app in Indianapolis, which is nearly an hour's drive from Bloomington, would provide my colleague with many more possible matches. The next problem would then become not how many people he could find in his search but how many would find him in *their* search. While it would give him access to more people, a wider metropolitan radius would also let him be viewed by many more potential partners. Mate search is both about seeing and being seen.

What's contradictory to common expectations is how these apps work. You don't just search and find a match; much of what goes into the formula is out of our control. Geography, population density, the parameters you put into an app, how the human brain searches and chooses, what you're looking for, what your potential mate is looking for — it's all part of the choreography.

That there are many factors at play, both digitally and neurocognitively, makes the search process profoundly more complex than most people appreciate. We can use this to our advantage, like my friend pausing in the bustling Indianapolis airport to broaden his app horizons. But what happens when those horizons feel intimidatingly broad? Dating apps make us face a "paradox of choice," just as we do when there are too many brands of toothpaste or detergent on the shelf, though choosing a potential mate is more momentous and psychologically daunting. We can react to this feeling, what psychologists call cognitive overload, by reframing our expectations from a state of searching for love to simply, and more accurately, choosing someone to chat with and see where things go.[15]

In the end, those "choices" are in many ways nothing more or less than an economic measurement, a way by which evolution has programmed us to look at what resources others have to offer versus what we have to offer, and whether we are displaying those resources in contexts of high or low competition. Are we beautiful? Funny? Intelligent? Are we a good listener? Do we have a nice family or a difficult one? And even: Are we financially stable (or better yet, wealthy)?

But most of all, context matters. Meaning, how well do we possess these alluring qualities *relative* to others competing for the same partners? Actual mate choice is a mix of what we want, our own preferences, and what we have to leverage, like our looks, resources, status, and personality. Courtship is a competition.[16]

Let's call the search for romance, sex, love, and intimacy what it is: a *mating market*. And this market can have some pretty harsh parameters.

THE ECONOMICS OF THE MEET MARKET

You know the old proverb "In the land of the blind, the one-eyed man is king"? That's one way to look at evolution: Natural and sexual selection is all about *relative* advantage.[17]

The metaphor of dating as a "mating market" describes the phenomenon in which individuals compete to make "bids" to members of their preferred gender(s) in order to secure an opportunity for a romantic or sexual partnership. Bidding, however, is a two-way process. While we seek certain qualities in romantic relationships, connection also depends upon our possessing what potential partners want. An individual with attractive traits has a strong bargaining hand and can be relatively choosy about what bids to accept. Less attractive traits mean a relatively weak bargaining hand. In the context of heterosexual partner preferences and mate choices, the mating market metaphor highlights how preferences of one sex/gender predict the bargaining hands of the other. This generalized principle holds for mate choices among people of all demographics, across age, gender, sexual orientation, and so forth.[18]

At the evolutionary level of analysis, we find two different mechanisms involved in mating psychology. First, there's what we *say* we want in a partner; second, there's whom we actually choose.

Our research shows that humans are dreadful at articulating what we want in a partner. This is why researchers will often design experimental tasks to test partner preferences rather than simply ask people. Most of us have exhaustive lists of what we're looking for in a partner. Does she like birdwatching? Does he know how to cook? Is this someone I can introduce to my grandparents? These preferences influence our choices in principle, but as my colleague Elizabeth Bruch at the University of Michigan points out in her research, sometimes there are simply limits on the choices we are able to make. We live and date in particular markets, which only include a set number of people in the first place. This is the heart of my previously mentioned colleague's airport dating app strategy.

But we don't always articulate other traits that have an impact on whom we choose. Women tend to prefer taller men, and men tend to prefer shorter women.[19] This is a consistent, albeit moderate, human partner preference that spans cultures and time and is probably related to gendered questions of protection and status between mates (especially, but not exclusively, in heterosexual pairings). Height is a deeply fascinating trait, in part because it's determined by both genetic and environmental factors. Your genetic blueprint dictates a range of possible heights you could land on, a minimum and a maximum, but within that range your early life experiences determine how tall you end up being, as everything from nutrition to physical activity, hormones, and even emotional support come into play.

Our mate choices can be thought of a little bit the same way as height: Potential partners have composite "mate values" that depend on our initial list of what we think we're looking for within the range of biological desires established by our evolved preferences. In one of her studies, Bruch found that people tend to engage in aspirational dating, punching about 25 percent above their own "weight" when chatting up someone on a dating app.[20] I have a hunch this is also likely true whether swiping on an app or mingling at the bar. Think of the old-fashioned stereotype of young straight women evaluating a male partner's status and wealth as indicators of social dominance and financial stability. Or the equally clichéd trope of the fifty-five-year-old man with the thirty-year-old wife who is leaps and bounds more conventionally attractive than he is. We are quick to judge partnerships that seem mismatched on the surface, but one's total mate value is much more complex than we imagine.

Different people bring a lot of different things to the table — personality and experiences, adventure and opportunity, safety and prosperity — and what strikes a chord for you may be different from what works for anyone else. "I'm with my partner because she's more attracted to me than anyone else in the world and can't get enough of my personality" is a nice fairy tale, but it usually isn't the whole story. If we think of potential mates as composites of various qualities, we can be a little more humane in our judgments of others' shortcomings.

There's good reason to not be too obsessive over needing to check off every quality on your preferred-partner checklist. The first is straight population demographics: If you are especially restrictive in the combination of traits you want, the reality is that there may be few to none who actually meet those strict criteria. A viral TikTok video in 2024 showed a variety of users — some jokingly and some seriously — declaring, "I'm looking for a man in finance, with a trust fund, six-foot-five, blue eyes." But when you run the statistics on finding a partner with those particular characteristics, as one user did, there appear to be only two single men in the entire country who fit the bill. When your standards are too strict, the odds are not in your favor.

The other reason for not being hyperfocused on a list of predetermined traits has to do with realistic outcomes. In a series of studies led by Daniel Conroy-Beam, mate preferences were not strongly related to actual relationship satisfaction — in other words, whether one's eventual romantic partner fulfilled all the preferences on their "list" didn't impact how satisfying the relationship was to be in.[21] Taken together, dating may be most productive and satisfying when we forget about our "perfect match" and focus instead on the people and opportunities to connect that are in front of us.

SHOWING YOUR HAND

Several years ago, after deciding to start dating again following a difficult breakup, I walked into a car showroom. Now, a central thing I know about the life cycle of a relationship is that priorities change over time, and our behaviors shift accordingly. At the beginning stage, looking to attract a mate, we want to put our best foot forward, even preen. Some use the term *peacocking* to describe our tendency to take whatever resources we have at our disposal and display them to a potential mate. This analogy is spot-on. What are flashy sports cars but fancy feathers that broadcast as symbols of style and wealth? If I'd had children, I might have chosen a utilitarian SUV, but I was looking to peacock a little, so I am only semi-embarrassed to report that I let a salesperson talk me into purchasing a zippy red convertible.

I went to fill out paperwork with the dealership's finance person, Tim. We got to talking, and he told me his girlfriend was attending law school at Indiana University. When I mentioned my work at the Kinsey Institute, he said she was really interested in sex education and law. Would I be willing to meet her? Of course! Tim and I became friends, but not long after the ink was dry on my new car's paperwork, he split up with the girlfriend. Soon after that, I joked over dinner with Tim that he'd need a new car now, too. He ended up with a black Mercedes coupe.

Not everyone chooses, or can afford, such an expensive type of peacocking, but we all find ways to shake our tailfeathers in order to win the dating game. Whether it's working out regularly to keep ourselves physically fit, choosing flattering or trendy outfits, or wearing our hair a certain way, we humans tend to focus on our outward appearance when we're looking to enter the mating market.

Mating market isn't just a figure of speech. As social psychologist Roy Baumeister has argued, thinking about human mating psychology in terms of economic theories can help illuminate some of the fundamental trade-offs we make when deciding whether to enter into new romantic and sexual relationships — or even whether to go out on a second date.[22]

In practical terms, these decisions are cost-benefit analyses just as much as the choices made around a poker table or in a grocery store; we are deciding how much time, money, and emotional energy we are willing to invest to ensure we are the ones who end up attaining a scarce and desirable resource.

These calculations would be a lot simpler, however, if we were only looking for one thing in a mate. What complicates the search is the fact that not only is mate choice multidimensional, but when we look at the research on mate preferences, there are ongoing debates about whether our desires are based in sex differences or in sex roles.[23] In other words, it's possible that the things that women find attractive in a male partner are broadly attractive to all or most women. Likewise, it may also be that what is attractive in males is attractive to everyone, or most everyone, who is attracted to men. If the latter were true, we would expect straight women and gay men to exhibit similar preferences in their male partners. Of course, as is the case with many of the most interesting parts of the research, it's likely that humans have evolved our preferences through some *combination* of sex/gender, sexual orientation, and targeting.[24] In one study, gay men on average preferred men who described themselves as masculine rather than feminine, but this preference was weaker among men who rated themselves as relatively feminine. Lesbians preferred women who described themselves as feminine-looking, but they did not discriminate against women who called themselves masculine-acting.[25]

Two recent studies have given us a window into the complex vehicle of mate choice by using enormous multinational samples. The first, conducted by biological anthropologist Virginia Vitzthum, who is a senior scientist emerita at the Kinsey Institute, was done in partnership with Clue, the menstrual period tracking app used by women all over the world. The researchers asked more than 64,000 women in 180 countries what they look for in a partner.[26] Old tropes would have us presume they're looking for wealth, status, and taller-than-average height, right? Nope. While those things are important to many, they weren't the traits at the top of most women's lists. The study found that the most important trait women are seeking in a mate is *kindness,* which, as researchers have noted, is key to enhancing those social bonds that allow us to thrive as individuals, partners, and a society. Next on the list was supportiveness, another trait central to creating intimate bonds. Physical attractiveness and financial security rated only moderately important to the women in the study, which included both cisgender women and transgender women.

Another large survey, this one by psychologist Richard Lippa in collaboration with BBC Internet, asked more than 200,000 people in fifty-three countries to rank mate preferences across a range of twenty-three categories of traits.[27] Across the board, the top five preferred traits were intelligence, humor, honesty, kindness, and overall good looks; women rated honesty and dependability higher than men did, and men rated facial attractiveness higher than women did. Heterosexual respondents were more likely than gay and lesbian participants to say that parenting ability was important in a partner, and that cultural factors had a bigger impact on people's rankings of character traits.

The large body of research on whom we choose points to a surprising punch line. Not only are our evolved mate preferences not necessarily predicated on an overt imperative to procreate, but they are much more complex than previously understood. Even if you're lacking in power, money, or a great pair of legs, if you've got a sense of humor and you're dependable, you still have plenty of desirable assets to stock up your stall at the meet market. Yes, physical attraction matters a great deal. But the qualities that satisfy our emotional and psychological needs matter just as much, if not more.

In studies led by my collaborator David Frederick, a social psychologist, we examined two large datasets — with a total of more than 25,000 participants — of heterosexual mate preferences and how they varied by gender, age, personal income, education, and appearance satisfaction.[28] We again found that, on average, men and women differed in which traits they indicated were "desirable" or "essential." Wealthier men, but not women, had stronger preferences for good-looking partners, and people who considered themselves good-looking preferred slender and attractive partners. Older men and women had weaker preferences for desirable partner traits, however, suggesting that with advancing age one's preferences may shift to more interpersonal factors.

We are currently running follow-up studies to look at these patterns across sexual orientations. What we do know from the limited research is that there are some differences in partner preferences among those in the LGBTQ+ community (in part because of different reproductive factors, and in part because of different social pressures and even stigma). Gay and lesbian couples are more likely to be in multiethnic relationships. They are more mobile, in terms of things like long-distance relationships or where and how far they're willing to move to be with a partner. But our research confirms that whether you're gay, straight, trans, old, young, or somewhere in between, *kindness* is a quality that is universally desired in a partner.

The data are unequivocal about one other quality humans prefer in a mate: We seek partners who can lead us to self-expansion. Universally, we find growth beyond the self, new experiences, and new ways of thinking to be an alluring aspect of partnership.[29] This novelty can also come by way of resource expenditure — say, a new partner surprises you with plane tickets to Paris or takes you on a hike through the woods to the top of an old fire tower and you're suddenly able to see a completely new view of a once-familiar forest.

The things people are looking for in a partner run the gamut from companionship to safety to sex, and the body and mind respond differently to environmental stimuli. This is why a person's sexual orientation, identity, and preference may be heterosexual, but in certain situations his or her behavior may be bisexual or even exclusively same-sex. These scenarios complicate outdated simple models of sexual orientation, suggesting that preferences are not static, and that what one wants — and ultimately what one experiences — depends on all sorts of internal and environmental factors. Both our biology and our ecology matter.

ALL ABOUT THE GENES?

But it's also important to note that not all our behaviors or decisions are related to preferences for reproduction. Sometimes we are driven just as definitively to find a partner we won't reproduce with, perhaps because we just don't want children, we prefer a partner of the same gender, we are medically unable to reproduce, or we are looking for someone past childbearing age. It's important to avoid the reductive, straw-man argument that all our decisions and behaviors are rooted in an irresistible biological drive to pass on our genes.

We may be governed by the same biological principles as all sexually reproducing species, but we are also a species with tremendous variety in what each of us desires at different stages of our lives. After all, we humans have the largest and most complex brain of all living primates, a brain that has tripled in size over the course of the past 2 million years of our evolution — and those big primate brains have equipped us to make facultative decisions (decisions that adaptively vary depending on different ecological conditions). Understanding this is, again, key to unlocking solutions to our intimate challenges and successes.[30]

This principle has played out in my personal life as well. One ex-girlfriend of mine had a reproductive complication that meant she would be unable to conceive a child without medical assistance. Discussions of this kind come up early in many relationships, especially as a couple is making decisions on the use of prophylactics. It threw a spotlight on the ultimate goals we each had for our relationship. We discussed whether she could or would try to harvest her eggs, in order to have them implanted in her uterus at a later time, or whether she might choose to adopt one day. We both recognized that there are so many children in the world in need of homes. Or could we just not have kids and focus on other experiences together? We eventually separated without reaching a resolution on this question, but the point is that her reproductive challenges did not disqualify her during the early stages of getting to know a partner.

In one delightful study, cognitive scientist Peter Todd and colleagues investigated the conundrum that while we say we want a mate who is like us (equally matched in terms of physical attractiveness, wealth, status, desire to become a parent, etc.), often what we *choose* is a partner who will provide advantages as we pass on our genes.[31] The scientists set up a speed-dating event, which gave them an environment more controlled than a bar but one that still allowed for spontaneous connection, and in advance of the event quizzed the college-age participants, all of whom were single and straight, at length about what they might be looking for in a mate, how they rated themselves in those same categories, and how many offers to meet again they would ideally want. As expected, men and women generally reported they were looking for someone equally matched. After the singles interacted, though, the researchers found that those evolutionary forces were at work shaping actual choice. The men tended to choose women based primarily on their physical attractiveness, while the women, who were much choosier than the men, selected men whose "total package" — their looks, yes, but also their personality, their seeming trustworthiness, their perceived ability to maintain a relationship — rivaled her own rating of her physical attractiveness.

This result is well aligned with the evolutionary predictions of what we call parental investment theory, in which we seek mates to balance the burden of reproduction.[32] Human females, with their limited numbers of eggs, will be the ones gestating, nursing, and often caring for offspring, and so have more to sacrifice in the process than males. To this end, women tend to choose men with greater wealth and status and who are somewhat older. Men, on the other hand, look for physical markers of fertility in women, preferring such traits as youth and waist-to-hip ratios that approximate an hourglass shape, which is, evolutionarily speaking, a sign of reproductive fitness. Across the animal kingdom, we see that the mate who bears the greater burden of reproduction is the choosier one. Humans, too. In one study of thousands of dating app users, men "liked" nearly 62 percent of women's profiles, while women "liked" 4.5 percent of men's profiles — a major sex/gender difference in openness to connecting with a potential partner on the order of men saying yes to two out of three profiles and women saying yes to two out of forty.[33]

But there's an added layer to this choosiness for us pair-bonding primates. When choosing a mate, we're also choosing a partner.[34] We don't just want sex for the purpose of reproduction.[35] We want to form a long-term bond as a foundation for reproduction, because we are intimate animals.

DEALBREAKERS

We are not prisoners of our biology. Individual differences in attraction are a major part of what determines the kind of partner(s) we choose. Based on an abundance of data, including that collected from over a decade of our research with Match, we know that people operate according to something like a hierarchy of preferences. These preferences are then more or less constrained by adaptations in our psychology, kind of like how a person's height is determined by environmental factors and biological ones working in concert.

The mate choice symphony is particularly cacophonous. Work by the social psychologist Eli Finkel has produced a fantastic theory he calls the "suffocation model" of marriage (he titled his book *The All-or-Nothing Marriage*).[36] This is the idea that modern couples expect to have all their needs — passionate love, caretaking, emotional support — met by a single partner, when humans historically have had an entire family network or village to rely on. Perhaps today we've come to expect our intimate partners to do too much, to be our everything. There's simply too much complexity in human need for this to work.

This "all-or-nothing" mindset may be why research shows that people give more weight to their dealbreakers — those undesirable traits or lifestyle choices that make us swipe left, literally and metaphorically — than deal*makers* when it comes to dating. People with higher "mate value" — that is, people with greater quality and a higher number of the qualities we look for in a partner — reject more traits as dealbreakers; people with less-restrictive dating strategies are less wary of them.

In our own studies, we've found that undesirable traits are more likely to break the deal in long-term relationships than in short-term uncommitted sex contexts. This makes sense. The more we invest in a relationship, the less tolerant we are of what we don't want. Overall, both men and women weighed dealbreakers more negatively than they weighed dealmakers positively, although this effect was stronger for women, especially in the context of a committed romantic relationship.[37]

Not all our mating preferences are intentionally developed, and some are highly dependent on the ever-shifting biases of our cultures.

For example, one study we conducted, led by Amanda Gesselman, revealed that in recent years sexual inexperience or "virginity" (the latter a loaded term that's been criticized by many feminist scholars) has acquired a stigma that can hamper finding a partner. Virginity, in some cultures, is now a social handicap for many men and women, while in earlier times, chastity was one of a woman's (and to a lesser extent a man's) more valuable commodities.[38]

Mate preferences and mate choices are central to our understanding of human relationships and sexuality in an evolutionary context. Searching for a partner is where the journey begins, and most people exercise relative preferences. But today, new cultural norms and the internet, especially the availability of dating apps, have fundamentally changed human courtship, shaping and reshaping the ways in which we seek out and connect with potential mates. We are a species in flux. With more opportunities, the search for a mate has simultaneously become easier than ever and harder than ever because the human brain hasn't evolved to choose from so many options. To truly build fulfilling, and lasting, intimate relationships, we must not only figure out what qualities we truly desire in a partner but also stay focused on the person in front of us long enough to learn if they're dancing to the same music.

Chapter Four

DATE

I led us both right up to the edge of the cliff.
As we laughed through the uncertainty of whether the leap we were about to make was more terrifying than fun, I may have puffed out my chest a little bit, like a gorilla displaying before a potential mate, as if to reassure her I could save her if she stumbled. In truth, my heart was thumping. I could almost hear the sounds of the sweltering jungle — monkey howls, bird choruses, water rushing — I'd encountered the last time I tried zip-lining in Monteverde, Costa Rica. Peering over the edge as sweat trickled down the back of my neck, I reminded myself we were only in the woods of Brown County, Indiana.

Still a cliff, though. Still planning to jump off it.

I had known the woman next to me for a while. She lived hundreds of miles away in another state, but after working together on a project earlier that year, we had been texting each other with increasing frequency. Our long-distance flirtation had been building over the last few months, but without any explicit plans to meet up. So I was surprised when she messaged to say she had plans to be in the Midwest, not far from Bloomington, and could stop by for a weekend visit. I immediately went to work planning the perfect date.

Enter the zip line.

She wanted to go first. Climbing into the harness, she looked over her shoulder at me, grinned, and then leaped out into the abyss. She screamed as she flew across the small lake below, her shrieks half real, half theatrical. I yelped just as loudly a few moments behind her, our cries echoing through the forest canopy like courting whippoorwills calling to one another among the trees.

There was a reason I had planned this activity, rather than dinner and a movie.

The goal of any first date, even the first few dates, is to get to know someone a bit better, to bring that initial chemistry to a simmer — and, if you're lucky, to a boil. I took my date zip-lining because it allowed us to walk and talk while experiencing something novel and thrilling together. There is a psychological principle, known as misattribution of arousal, that we might put to good use in the dating arena; I knew what we were doing could arouse certain inescapable feelings of excitement, and that those could then work to my benefit in the business of trying to impress.[1] Zip-lining is a thrilling experience, and my plan was that by doing it together, the very notion of being around me would be perceived as thrilling, too.

Yes, this is how sexologists approach first dates.

BIOHACKING CLOSENESS

The modern concept of dating, in which two people engage in an unchaperoned social activity to assess attraction and compatibility, has really only been around for a little over a century; in fact, the term *date* was first used in this context in 1896 by a columnist for the *Chicago Standard,* writing about a jealous clerk who had confronted his girlfriend about "other boys fillin' all my dates."[2] The Western idea of "courtship" emerged during the eighteenth century, but back then, it was largely a family affair, governed by strict rules and rituals including kin approval and an exchange of resources between families prior to marriage, for the express goal of finding a match that would lead to a mutually beneficial marriage.[3]

In the twenty-first century, dating is a much more fluid and personal human interaction. The reasons we date are myriad and complex, and it isn't necessarily about finding our forever person. We date to find out if we're attracted to someone and to the multiple aspects of who they are. We date for fun and companionship, for romance, to discover if we have sexual chemistry, or just to not be alone.

Dating is where the tension between our primal motivation for sexual exploration collides with the deep human need for intimacy.

Misattribution of arousal, the phenomenon I was tapping into on that first date, was first assessed in the context of interpersonal attraction in a now-classic 1974 study by psychologists Donald Dutton and Arthur Aron.[4] In their experiment, one group of men walked across a somewhat scary, rickety-feeling suspension bridge, while a control group of men walked across a sturdy bridge. Afterward, when both groups were greeted by a young female research assistant who gave additional instructions along with her phone number, ostensibly in case they had any follow-up questions, the men who walked across the unsteady bridge were far more likely to use that phone number to contact the woman afterward. Half of the men from the suspension bridge called, whereas only one out of every eight from the stable bridge did.

What subsequent replications of this type of experiment have shown is that a physiological arousal response from novelty, anxiety, or even fear is often ambiguously interpreted by the brain. Meaning that we sometimes attribute a feeling to a stimulus different from its actual cause. For the men in Dutton and Aron's experiment, fear of falling off that rickety bridge got translated into arousal, then attributed to excitement and even romantic/sexual interest in someone nearby.

In the context of my date, wielding this knowledge was not nearly as sneaky as it might sound. My zip-lining partner, a behavioral scientist herself, knew about the phenomenon, too. But newer studies have demonstrated that the misattribution effect still plays out even when the subjects are aware of the actual source of arousal. Zip-lining is also just a really fun thing to do on a date, in my opinion: a great way to gauge a potential partner's level of sensation-seeking (along with other facets of their personality) and hopefully ignite a spark of connection that might very well lead to a long-term partnership.

We tend to put a lot of pressure on first dates. But the research suggests that *second* (or third or fourth) dates matter as much as first dates. In our 2022 Singles in America study with Match, we found that over 70 percent of people say they've fallen in love with someone they knew first as a friend or colleague, and that nearly half (49 percent; 44 percent of men and 52 percent of women) say they fell in love with someone they didn't initially think it could happen with.

I like to think of dating as a two-step; we put our best foot forward to make a good impression, followed by our less attractive foot to facilitate the growth of mutual trust. Maybe the first foot is perfectly manicured, while the second has chipped nail polish and that weird pinky toe we broke in high school field hockey practice. But both feet are necessary — we need both to walk, then run, then dance with our prospective partner. When we choose a partner, we look for someone we find attractive (physically, behaviorally, socially, emotionally, and spiritually), but also someone who will reveal themselves in a way that allows us to assess their character accurately.

That second foot is important because if we sense someone is deceitful, we won't trust or feel safe with them, both clearly nonstarters for a stable romance. My Kinsey Institute colleague Stephen Porges describes this pattern in his widely regarded polyvagal theory, which explains that our nervous systems detect physical and social environments, then detect whether situations are safe or dangerous using an evolved social engagement system.[5] According to Porges, our subconscious nervous system assesses the safety of our environment through a process called neuroception.[6] When our bodies feel safe, we can engage, empathize, and connect interpersonally, whereas the opposite state — fear — induces biological threat and flight responses. Our bodies and brains, in other words, have evolved so that feelings of close connection require a backdrop of safety. In the absence of trust, forming lasting pair bonds can be difficult, not to mention emotionally painful.

In threatening situations, our nervous system is called into action, priming us to flee and focus on survival. So, while a little bit of novelty and risk (say, the rush of leaping off a cliff strapped into a harness attached to a zip line) might lead to misattribution of arousal and romantic interest, too much adrenaline can have the opposite effect, making it difficult for our brains to make meaningful interpersonal connections. In fact, when people do form close bonds amid dangerous environments, it's typically because the other person is a source of emotional or physical safety. When our partners provide that to us, we are better able to weather the storms of our lives.

ACCELERATING CLOSENESS

In 2015, the *New York Times* ran a "Modern Love" column that might have made it into your inbox. The title, "To Fall in Love with Anyone, Do This," apparently tapped into some of our collective anxieties and quickly went viral.[7] In the essay, writer Mandy Len Catron tells the story of using a set of increasingly intimate questions to get to know a potential romantic partner. The thirty-six questions, actually developed many years prior as an experimental research tool by Arthur Aron and colleagues, are designed to accelerate close connection between two people.[8] They start with "easy" ones, like *Would you like to be famous? In what way?* (#2), but lead inexorably to *If you were to die this evening with no opportunity to communicate with anyone, what would you most regret not having told someone? Why haven't you told them yet?* (#33).

The purpose of this task was to help the pairs develop a close bond in a short period of time, by accelerating through the moments of connection that would otherwise naturally occur as a friendship or relationship develops. As the researchers wrote in their landmark 1997 academic article, "One key pattern associated with the development of a close relationship among peers is sustained, escalating, reciprocal, personalistic self-disclosure." The *New York Times* writer who tried this test nearly twenty years after it was developed answered honestly and thoughtfully, then...fell in love.

That writer's experience isn't dissimilar from what comes of advice dispensed by relationship experts John Gottman and Julie Schwartz Gottman. Their book *Eight Dates: Essential Conversations for a Lifetime of Love* walks readers through a series of conversations on everything from finances to sex, all of them designed to help couples enact vulnerability and therefore connect more deeply.[9] These dating tips are consistent with what we know about the psychology and biology of close relationships. Dating is the process of getting to know someone, and disclosing yourself to them is essential to establishing close bonds — and eventually, possibly, feelings of romantic love.

The proliferation of technologies has made it easier than ever to find potential dates, but at some point you have to actually meet the person, and dating is the process of engaging your brain with someone else's, accumulating information, and using all your senses to narrow in on whether that person has what you truly want in a partner. It's all trial and error, and what science tells us is that going on those dates is important to cultivating intimacy.

A friend of mine had been divorced from her husband of a decade for about three years when she finally downloaded a couple of dating apps. "It's just so exhausting to even know where to start," she'd say to me anytime I saw her. "The process of learning about someone new and sharing things about yourself...ugh." Getting to know someone can feel daunting, but she persisted, if for no other reason than to meet people in her city. Her first week of swiping netted a date with an attractive lapsed Catholic with a long history of exploring psychedelics, a high school history teacher who coached powderpuff football in a pink tutu, and a fisheries biologist who'd lost his mother as a young man, around the same age she'd lost her father. When I spoke to her next, she was beaming — "Justin! People are just *so interesting*!" — even though none of those dates had generated a romantic spark.

My friend's delight points to a central truth about dating: It is a highly efficient process by which we learn about other people. Because humans form preferential social relationships, most of our interactions are with people we don't necessarily know very well. You might see a neighbor or colleague every day for ten years and never know that person as well as someone you've just met but with whom you've just shared your happiest childhood memories. Exercises like Aron's thirty-six questions are designed to reveal the most interesting, truest things about a person, and self-disclosure, trust building, and mutual appreciation accelerate closeness. If you're doing it with someone you're attracted to and if you're psychologically available, it can also be a shortcut to building romantic feelings.

The more you know about a person, the more likely you are to find common ground. Let's say you're a vegan whose passion is rock climbing, but you meet someone for whom the idea of a perfect date is a steak dinner followed by a cozy night in reading on the couch. You very well could be a good match for a whole host of reasons, but if outdoor sports and environmental activism are your primary interests, your predetermined dealmakers, you're probably not going to ask the meat-eating homebody out for a second date unless you learn a lot more about them. As we get to know each other we can sometimes find unexpected opportunities to connect with and be attracted to each other.

The same tools that help two strangers accelerate intimacy can also help couples maintain an intimate connection as their relationship matures over time. Dating shouldn't be a process that ends when commitment begins. My research has convinced me that the best way to maintain intimacy in a relationship is to continue to date your partner throughout the entirety of your relationship. Yet, finding the time to reaffirm those intimate bonds can easily get lost in the shuffle of our busy lives, especially in cases where people have very different social networks or work schedules or live in different geographical regions. One long-distance relationship strategy I've personally employed is to watch or stream a TV show with a partner either synchronously over video chat or while live-texting each other. Watching the same reality TV show from your respective homes may not seem like a bonding activity, but research studies have demonstrated that sharing media or entertainment is associated with greater relationship quality, especially in cases where couples don't have common social circles. It gives us something to talk about, and a jumping-off point for exploring each other and ourselves within the partnership. What storyline or character did you find funny, interesting, or upsetting, and why? Shared activities like watching a common television show can help promote high relationship satisfaction by providing an opportunity to connect while exploring each other's likes and dislikes.

Studies have shown the benefits of mutual ritual: The things we do over and over with a partner can help us build security and trust.[10] But the flip side of such routine is the potential for boredom and lethargy. Satisfying relationships have consistent patterns and rituals while also making room for curiosity and novelty.[11] There are a lot of ways to accomplish this balance, and it often doesn't take much. For example, during the pandemic, a friend of mine and her partner got in the habit of taking a daily walk through their neighborhood, phones in pockets, just walking and talking, together. They would observe the changing seasons through their street's trees, note any new holiday decorations, point out interesting landscaping. As new homeowners, their walks gave them ideas for how to continue to make their house feel like a home. But over time, my friend told me, the ritual got a bit stale. As the months of quarantine went on, she discovered that for the walks to be both pleasurable and useful, they needed variety: Things didn't change enough on their street to make it interesting, day after day. So they began altering their route ever so slightly, walking one block farther north, west, or east. Sometimes, if they felt they'd seen all they could see, they'd get in the car and drive to another neighborhood. The routine required curiosity and variety to remain satisfying.

For our relationships to work, we have to stay curious: about ourselves, about our partner, and about the two of us together. And we have to continue to pay attention through all the changes.

After Jen and Dave got married and had their first child, it got harder for them to find time for the activities, like road trips and mountain biking adventures, that had brought them closer together when they were first dating. So my parents would insist on watching the baby so they could go on an occasional date night. As sleep-deprived new parents, Jen and Dave would usually just go out for dinner or sometimes a movie, but those nights out were an infusion of much-needed bonding as they navigated the transition from pair bond to parenthood.

Whether it's two people on a first date or a married couple out on a date night, the important thing is to engage in activities that allow each partner to focus on the other. What we're really doing when we date is carving out time to share an experience with each other through a common activity.

Garden together, go for a walk. Take a quick zip line adventure, maybe.

THE DATING GAME

Ever think too much about the way you're walking on a date, and suddenly you're unsure which foot swings with which arm? Me too. Some of us are more or less doomed, it seems, to spend much of first dates saying and doing awkward things. But there is hope. In one study, we found that 61 percent of men and 70 percent of women are more excited by a second date than by a first, and nearly half of men and women believe that a person with whom they had an okay first date can grow on them with more time.

The inevitable awkwardness of a first date is why I'm a big advocate of video chats and video dates to gauge level of interest before taking the leap to meeting in person. Whether messaging with a potential partner on a dating app or texting with someone new prior to going on a date together, many people feel pressured by the jump from sending brief messages to meeting in person. During the 2020 COVID lockdowns, we saw a sudden rise in video dating. In our national survey of adults in the United States, Helen Fisher, Amanda Gesselman, and I found that 68 percent of singles had used a video date to determine if they would go on a date with someone in real life, with seven out of ten saying they would video-date again. Of those singles who utilized video dating during the pandemic, half of participants even indicated they fell in love during video dates.

Video chats and video dates provide visual and auditory cues, including both verbal and nonverbal body language communication, that go beyond what we can get from a text or direct message.[12] Our data show that people are using this emerging technology in a multitude of ways — some meet over video for a brief conversation and "vibe check," while others engage in an entire date, including a meal, drinks, and even games. There's less social pressure, less time and financial investment, and an added sense of safety. For those venturing into the uncharted waters of digital dating, it's important to remember that as far as our evolved mating systems are concerned, this is as much of a two-way vetting process as an in-person date.[13]

Over the last couple of decades, as we have begun to navigate this next stage in the evolution of human courtship, an industry of professional services has sprung up to provide guidance on how to date effectively online. In fact, some dating apps now offer sessions with a "dating coach" as part of one's subscription, and many of the top professional matchmakers I know actually require some coaching for their clients. In order to keep their success rates high, many of these companies and boutique services want to ensure that their clients know how to date in a way that is fun for both parties and produces desired results. This is preferable to having someone offend other clients, or an unsatisfied customer later blaming the app or matchmaker when they can't quite land a match.

Dating coaches tend to disagree on whether it's best to keep first dates short or to demonstrate investment with a bigger time commitment. My theory as to why there's no consensus on this issue is that courtship is a profoundly personal and variable experience. For one person an ideal first date might be a leisurely two-hour dinner, while for another it might be having a chat while grabbing coffee on their lunch break. What we want out of a date depends entirely on the investments we're willing to make in that person at that time.

There are pros and cons to either strategy, but limiting first dates to the length of time it takes to drink a latte can facilitate a low-risk introduction and reduce the stress that would come with sitting through a longer activity if the chemistry isn't there. But you don't want to keep it so brief that you don't have time to get into the flow of conversation; showing that you're engaged in culture and real-world topics has been found to increase the chance of scoring a second date by 91 percent.

As we date and get to know our partners, the introduction of sexual behaviors is a natural progression. The first sexual encounter with a new partner can be an expression of how we feel, a gesture indicating that we want to pull the other close both emotionally and physically. Or it can be aspirational, an indicator of what one or both imagine as a possible future. For some, it can be pure physical gratification. Regardless of our motivations, our bodies and minds become uniquely engaged, which for many is more than just a straightforward matter of pleasure. We learn a whole host of new things about our partners when engaging in sexual activity together.

During the dating stage of a new relationship, the timing of a first sexual experience tends to vary considerably, from those who have sex as part of their initial dating screener to those who follow a "not before three dates" rule and those who wait many months or until a commitment of exclusivity. As one great example of what sexologists call the hierarchical reordering of sexual behavior, we've seen that, relative to their parents' generation, young adults in North America today are more likely to engage in their first oral sex experience before ever having penetrative intercourse sex.[14] Just a few decades ago, oral sex was thought of as something more vulnerable than intercourse that you might only do with a partner you've gotten quite close to. But for many young people today, oral sex is viewed as less intimate than intercourse and might even occur more casually and be viewed as less risky.

Regardless of how it manifests, sexual behaviors are, for many, a part of the courtship process — an important stage between the first spark of chemistry and the possibility of forging enduring bonds. And despite the wide variety of roles sex can play in dating, there are some patterns that we can learn from.

SLOW LOVE

Sexual activity is often the start of more enduring connections between a pair, though certainly not always. People have long had their share of one-night stands, but for those working toward a romantic relationship, sexual activity will almost inevitably become a part of that process. Just as kissing can be vitally important to the human animal in terms of discovering whether a potential mate's touch and scent are compatible, sex is where we explore how our partner feels, smells, tastes, and what their other relational tendencies are. Do they like to cuddle? Do they compromise? Are they creative, caring, and attentive? Playful? Funny? We learn an awful lot of biological, psychological, and social information about someone during sexual activity. Of course, this is true for dating writ large; what we eventually learn when we spend time with a potential partner is whether we can share our lives with that person, whether for a handful of months or for multiple decades.

One global trend we've observed is that people today are taking more time with this process of getting to know their partners than they have in previous generations.[15] While they may be having sex earlier in relationships, they're also being more careful with their hearts and are more reluctant to be vulnerable. Helen Fisher and I have studied this phenomenon, which she calls "slow love," in our Match study. Some data even suggest that Generation Z and millennials, as compared to their parents, are dating less, having less sex, and marrying much later.

Changes in what researchers call "coital frequency" (how many times a person has partnered sexual intercourse per week) may be an indicator of changes in relational interactions. According to findings drawn from the General Social Survey, which is an important long-term longitudinal nationally representative study of how people in the United States behave, from the late 1990s to 2014, respondents went from having sex sixty-two times per year on average to fifty-four. But this so-called sex recession was not uniform among participants. The drop seems to be mostly among younger people and married adults. But there may be a more nuanced story embedded in the data: It is entirely plausible that this decline in frequency is an indicator not that young people today are allergic to sexual intimacy but rather that they value it even more.[16]

The rise of "hookup culture" over the last half century is evidence of this change in behavior, if counterintuitively.[17] Our Match data show that roughly two out of five people have had sex with a potential partner before the first date — what Fisher calls the "sex interview" — in part because trying out sex with a relatively unknown new partner is one way to determine whether they are worth the investment of getting to know via dating.[18] In other words, the slow-love process means that young people are taking time to really get to know a new partner, exercising caution before making a serious commitment. As a result, the average age of first marriage for people in much of the developed world is later than in past generations, and this isn't a bad thing: Relationships that take more time to nurture also appear to be more stable over the long haul. As Fisher and I write in a chapter for *The New Psychology of Love:* "Today's singles appear to want to know everything about a potential partner before they invest their time, money, and energy into initiating a formal commitment.... Where marriage was traditionally the beginning of a partnership, today it is the finale."[19]

What might appear to be a signal of disconnection or aloofness in today's singles could in fact be caution. Hookups and casual sex are, in many ways, quite risky: They involve going to someone else's house, often removing much of your clothing, and showing parts of yourself that most people don't see. Some are even vulnerable enough to fall asleep next to these relative strangers. I am not a moralist about these decisions, but I am curious about the behavioral risks we do and don't take. That so many millions have and will do this with people they barely know might seem surprising, but then again, we are a sexual species.

People generally find some degree of risk-taking sexually attractive (with some caveats for the type and severity of risk); similarly, people will often engage in risk-taking to be close to those they are attracted to.[20] One interesting line of research has shown that people are less likely to use condoms or other forms of contraceptives if they are very physically attracted to their partners, and likewise if they're committing infidelity.[21] It's possible that this is because risk-taking and novelty-seeking drive up the brain's dopamine activity and attention to the very thing that is risky. I've argued that pair bonds help us adapt to the uncertainty of the world around us, but that uncertainty can also be alluring and intoxicating.

Why, then, are so many people so slow to reveal themselves in less physical ways when they're dating? Why is it scarier to tell a date about your deepest fears than it is to hop into bed with them? It is because the motivational systems for sex and intimacy are not exactly the same. In a world in which social stigma about sex has fallen by the wayside — thankfully! — the old barriers are no longer there to protect us from showing too much of ourselves. The caution that so many show when they are dating casually or hooking up — slow to self-reveal, slow to define the relationship, slow to commit — isn't an indication that they don't take love seriously. *It's the opposite.* They are cautious *because* they take love seriously. Even if they are less cautious than previous generations about having sex outside of traditionally defined relationships, they are being cautious about their emotional vulnerability. They might be quicker to have sex, but slower to become emotionally connected and bonded.[22]

Our sexual desires, need for intimate connection, and appetite for risk can combine and collide in awkward ways. That's one of the things that can make dating so unpleasant and our behavior while we're on the dating scene so unpredictable, even embarrassing. When I was single and looking, I once found myself jogging on a treadmill at a gym when a disarmingly gorgeous acquaintance hopped onto the machine next to mine. We said our hellos, and what happened next was cognitively blurry to me at the time, though I later understood it completely as an ancestral need to try to impress this person with my athleticism. I started running my daily 3K; then, almost as if I weren't in control of my own right hand at all, my fingers kept drifting toward the speed-increase button. Before I knew it, I was sprinting faster than I had since junior high school, while she trotted along on the next treadmill over and tried to make casual conversation. I am 100 percent sure she didn't notice the miles-per-hour designation on my machine, but she definitely noticed when I hit the stop button and almost fell off the rails, gasping for breath.

Later, as we walked out together chatting, an athletic-looking man joined us. "Oh, Justin! Meet my boyfriend, Tom!" she said. In the moment, I was too exhausted to experience the full range of my defeat. I just knew that my biological drives had collided, disastrously, with my understanding of the social code. And that my calves were sore.

IT'S COMPLICATED

Dating is an intricately choreographed push and pull — enticing with what we have to offer, pursuing what we want, and responding to what we like. But courtship rituals also change with time and place.

Changes in the ways we meet potential romantic partners have, unfortunately, ushered in new forms of bad behavior and novel questions about the nature of consent. In one of our national studies of singles, we found that roughly one in five participants reported sending a sext (i.e., a sexually explicit image sent via text message), and 28 percent had received them. And although nearly three-quarters of participants reported worry that whoever they'd graced with a sext might forward it, around 23 percent had shared sexts they received with others, on average with more than three friends.[23]

In a second, more recent study on sexting led by my Kinsey Institute colleague Alexa Marcotte, we also found that sexual images, and specifically the notorious "dick pic," may now play an even more frequent appearance in early courtship and pre-dating behaviors. In our sample of more than 4,200 single adults, we found that nearly half of women across sexual orientations and ages had received an unsolicited male genital image, and so had two-thirds of gay and bisexual men.[24]

Some of these new patterns are unsavory, but they are perhaps not all that surprising. There has always been misalignment of intent in courtship; prospective partners have always been in danger of misinterpreting signals (or lack thereof). Think of it this way: Fifty years ago, a man might misread a situation with a woman he had recently met as a romantic flirtation, then make the embarrassing decision to send her unwelcome flowers or gifts. But in today's currency these overtures can be more extreme than receiving unwanted flowers, as an unsolicited nude photo seems to ratchet up unwanted attention that feels intense, presumptuous, and, in some cases, threatening. This difference is possible because we now have a technology that facilitates taking more risks from behind a screen and allows us to send images instantly, to many people at once, at very little cost. And when we have never actually met the person who is being showered with unwelcome attention, the social consequences are low if the overtures are rejected.

The other difference is that flowers are generally accepted across genders as pretty and pleasurable, and sending them is generally considered a socially acceptable act of courtship. This is not the case for photographs of the male anatomy. In fact, studies show a fundamental disconnect between how heterosexual men and women view this practice. Men — regardless of orientation — tend to think sending a photo of their genitals is, if not exactly wholesome, at least fun and flirty, whereas heterosexual women, on the whole, find it gross.

The problem is in the betting odds. Men keep sending these pictures because every so often it *works*. Someone *is* into it. And since there is so little cost — monetary or social — to spamming the dating networks, they continue to do it.

So is the genital pic a courtship ritual? It certainly is seen that way by some people. And within committed relationships, it has become a tool for intimacy. Studies show that sending nude photos is actually more common *within* committed relationships than outside of them.[25] Remember, we are visual creatures. And with the rise of internet dating and long-distance dating, it's only natural that we reach for our phones to help keep the spark alive: after all, we're using them all the time, anyway.

Unsolicited nudes are only one of many new concerns to arise in the age of internet dating. Another is the question of authentic identity, or catfishing.[26] Though Cyrano de Bergerac–type deception is not entirely new historically, it has never been so easy for a person to pretend to be someone else in order to lure a mate. By now, we're all familiar with stories like the one told to me by a young man who had struggled with dating until he met someone online who he thought was his soulmate. Over the course of several months, Peter had developed a real emotional connection to this woman. The two of them messaged multiple times a day, his trust and affection growing as they shared personal stories and dreams. He began to imagine a future together, and was devastated when the love of his life turned out not to be a twenty-five-year-old yoga instructor in Duluth but a sixteen-year-old having a prank contest with her friends.

But what about the feelings he had for the yoga instructor? Could those still have been real love? I think they probably were. It's possible to become deeply attracted and even attached to the idea of a person, even if the actual person turns out to be a fake. This is especially true when we see images and have intimate, real-time communications that allow us to build a fantasy of who we think that person is and might be. The feelings that grow when getting to know someone can be honest and intense, even if they're based on an illusion of trust that ultimately is absent.

I would be remiss not to point out that while catfishing is a real risk, it is still relatively rare. All mating contexts come with unique risks, and balance those potential risks with potential rewards. For so many people internet and app dating is an opportunity to hit a romantic jackpot. Back in the early days of online dating, one close friend of mine met a man online, and when they finally got together in person at a café, she learned that he drove the exact same car as her, down to the model and color. My other friends and I, observing gleefully from the diner across the street, took this as a sign either that she was destined to be with him or that he was a serial killer and she would be his next victim.* Several years later I am happy to report that it was destiny, and that I was honored to officiate at their beautiful outdoor country wedding in upstate New York.

YOU AND ONLY YOU

Whatever form it takes, courtship is an essential process for the human animal invested in transitioning from being single to being paired. But how do two humans decide to share their lives, their passions, fears, and innermost selves, exclusively with each other and no one else? What does it look like when we say "You; yes, you; yes, you and only you"? And why do so many of us want it to be "just you and only you," often to the point of mutual possessiveness?

One classic "tell": We give each other pet names. In our research, my colleagues and I found that more couples give each other these names than you might think, and many of our participants, across all demographics, also engaged in "romantic baby talk" (or what we called "loverese") in new relationships, in some cases averaging up to ten minutes of loverese per hour spent together.[27] Despite them sounding ridiculous to anyone else within earshot, it can be important to create these special things we say or do only with our partner; in the collaborative relationship study of nearly 100,000 U.S. participants that is behind *The Normal Bar*, a book by Chrisanna Northrup and sociologists Pepper Schwartz and James Witte, nearly two-thirds said they used pet

---

* For the record, it's a good idea to have your friends help vet the person you're dating, and definitely to make sure they know where you are on a first or blind date.

names in their relationship, and roughly three-quarters of those who were "very happy" in their relationship used pet names with their partner.[28]

On a more granular level, there are so many interesting questions about how we signal our exclusivity, especially in the context of social technology and online dating. At what point in the relationship does a couple start to post photos together on Instagram? Share passwords to devices? Take down your dating app profiles? Stop talking to other potential suitors?

Whereas not too long ago a couple becoming "Facebook official" was a milestone, today we're now seeing a much more subtle indication of status across social media.[29] How often do two people post photos of themselves together on Instagram?[30] How many heart emojis do they use in the comments on each other's posts? The story of a relationship is now told through multiple data points across platforms and contexts. "Facebook official" is also no longer a reliable indicator of relationship status, as many folks use the platform to indicate affinity with a best friend or as a joke. In fact, I was "Facebook married" to a college friend for years before a new boyfriend of hers pointed it out. Understandably, he was a little displeased.

By understanding why we behave the way we do when dating, perhaps we can begin to tease apart some of that complexity, to both achieve our goals and enjoy the process. The most pertinent thing we can take away from our biological need to move from the *search* stage of the journey to the *date* stage is that at some point we have to actually get to know people. But if we can harness an understanding of our biology to hack the dating process and make it fun, it will be easier to choose someone to actually *mate* with. And, despite our culture's glorification of "happily ever after," we would do well to avoid measuring the quality of a relationship by its longevity.

Take my date in Brown County. Even though we zip-lined that romantic weekend into a deeper relationship, it eventually ended rather painfully. But I still genuinely regard the experience as a success. The relationship helped me realize something about love I had previously understood intellectually but not in practice. One of the harder truths of this science is that sometimes love is not enough. Philosophers, poets, and musicians can muse about love lifting us up, keeping us warm, and solving famine and war — and I certainly believe there are instances when it can — but the simultaneous reality for many is that relationships are more complicated than that. It's a perplexing truth that not all relationships are characterized by passionate love, and not all romances are sustainable relationships — but sometimes, with a bit of science and luck, we can enjoy that indulgent combination of having it all.

A few years later, I would find myself back in Brown County, on another first date — this time for logistical reasons. Little Nashville, Indiana, is a quaint midwestern town nestled between two state parks and surrounded by several covered bridges. With its charming shops and thriving arts community, it was the perfect spot to meet up for brunch on a spring weekend. It also happens to be halfway between where Michelle and I lived. We ate at a country café and then took a long leisurely stroll through the town, poking in and out of little candle and fudge shops. She bought a set of hand towels for her kitchen; one had the Eiffel Tower embroidered on it, and we joked about taking a trip to Paris together one day.

Despite an undeniable spark of attraction, we were both at a point where we were quite happily settled in our single lives — at one point, Michelle even said as much. We weren't dating with the expectation that we would find someone who was extraordinary. Yet, as so often happens in the pursuit of love and intimacy, that was exactly what we found. At the time, I was traveling a lot for my work with the Kinsey Institute, and she was focused on her own career. We lived over an hour's drive from each other, so meeting up on free weekends required a degree of planning that, for another potential partner, and perhaps even at another point in our lives, might have resulted in that initial spark petering out. We were content with a slow build, however, and over the next four or five months, we shared new experiences, laughter, flirtation, and more laughter. And our worlds became better with each other in them.

Dating is how we get out of the endless searching. It's the process of finding the partner with whom we want to build a relationship and ultimately even share a life. It's about connecting with another person and gathering the information we need to say "I choose *you*" — which for so many of us is the next step in the relationship life cycle.

Chapter Five

MATE

One aspect of my job I rarely discuss in public is that it comes with an element of risk. The vast majority of people I encounter out in the world and online are supportive of the Kinsey Institute. But, as with any science that pushes at the boundaries of what we know and accept about the human condition, there are also those who are angered by the work we do.[1] I'm not sure I fully appreciated this reality, or the ramifications for my personal life, until I woke up one morning to find myself the target of a malicious internet troll and their goblin comrades.

I was on day three of a COVID fever that had laid me out, so when I woke up at 5:00 a.m. to a barrage of increasingly alarming emails, texts, and notifications, my first instinct was to reach over, wake up Michelle, whose symptoms weren't quite as bad, and ask her thoughts. I was both feverish and half asleep, so I don't remember exactly what she said. What I do remember is the soothing sound of her voice, the warmth of her body next to mine, and experiencing a moment of clarity, a *knowing*, that whatever storm lay ahead, the person lying beside me would weather it with me. I felt safe and supported enough to put my phone down and go back to sleep. The other thing I remember is that just before drifting back into a fever dream, I made a mental note to install a proper security system in my home.

It had been a year and a half since that first date in little Nashville.

Our "slow love" had progressed beyond attraction and enjoying each other's company, beyond sharing our hopes and dreams, beyond meeting each other's friends and family, and even beyond discussions of marriage and children. I'd bought the engagement ring and had started planning a proposal. I already knew (to paraphrase a line from *Grey's Anatomy*) that Michelle was "my person." And yet, in this moment, I experienced that knowing again, perhaps even more powerfully than before, because this time it came with the realization that if we were to merge our lives, I needed to take steps to protect her from very real threats that in the past would have impacted only my own home and life. Both gradually and all at once, my world had expanded to include another human to whom I was now connected, for better and for worse.

In our most intense romantic relationships, there comes a point when our partner becomes a part of us. We start to think not just of *you* and *me* but of *us* and *we*. Our mindset shifts to shared goals beyond the individual, what behavioral scientists call "superordinate goals."[2] But social psychologist Arthur Aron and clinical psychologist Elaine Aron proposed that, rather than losing ourselves in the shadow of this other person, we are motivated to form these bonds as a way to enhance our own sense of self and self-efficacy. If we think of the grand evolutionary adventure of pair bonding as how humanity has adapted to respond to uncertainty, then "expansion of the self" is the process by which we become greater than the sum of our parts.[3]

It's also the process by which we transition from falling in love to *being* in love.

This is a psychically big shift for people to make; it's also one that's not neatly defined. If we look at the academic literature, we can find research on courtship, mate choice, dating, cohabiting, relationship maintenance. But the research hasn't done a great job of pulling out all those micro interactions that exist somewhere in the nebulous space between "just dating" and "till death do us part." Things like deleting dating profiles, cutting off a "plan B" potential partner who may have been waiting in the wings, taking a vacation together, and changing our eating habits to accommodate each other's diets.

To measure the intimacy experienced in pair-bonded relationships, Aron and colleagues developed the Inclusion of the Other in the Self (IOS) scale, in which respondents are shown seven pairs of circles ranging from just touching to almost completely overlapping and asked the question "Which picture best describes your relationship with your romantic partner?"[4] As couples enter this "mate" stage of a relationship, those two circles tend to begin to overlap more and more. As our lives and sense of self start to intertwine, we begin to see a little piece of each other in how we think, speak, and act. It's not just that we're attracted to each other, or even "in love" with each other; it's that we're pulling each other close.

We're also opening our hearts to risk.

When I made that mental note to install a security system in my home, it wasn't about my own safety — it was largely about Michelle's. As this intimate bond formed between us, worrying about her wellbeing had become a part of the way I think and live.[5]

When we get to that point where we decide "I choose you," the person we're dating stops being a passenger on our journey and becomes a co-pilot with whom we can navigate life's triumphs and trials. We want partners who widen our lens and help us experience the world in new ways, who bring us pleasure and novelty not just in the sexual sense, but also in terms of how we experience life. The excitement of novelty, almost by definition, is the introduction of uncertainty — and as I've proposed, one cornerstone of our evolved pair bonds is finding a desired partner to weather all forms of uncertainty with, together. In many ways, this psychological overlap with our mate is how we attain cognitive resolution between our sexual impulses and our primal need for intimacy.

BOTH FEET IN

If dating is the process of narrowing in on that one person from among the crowd, this next phase of the relationship life cycle is where we decide to put both feet in: "I'm going to invest in the person in front of me, and everything that comes with that. I'm choosing all this person's wonderful qualities, choosing the joy and the passion we've shared and have the potential to share, but I'm also choosing all their quirks and flaws — all the things I don't like about them and all the things that aren't perfect about our relationship."

This stage of modern mating, particularly in the context of the "slow love" pattern we see in the developed world today, is one that's getting longer and longer.[6] More people than ever, particularly young adults, are spending longer periods of time getting to know everything about their mate before deciding if they are "the one": everything from how they fight to how they make up, manage stress, and interact with our family, friends, and co-workers. This is a profound shift for most of us. Although we might be conscious of some aspects of the processes that bond one human being to another, much of this selection is driven by innate impulses — many of which we share with other animal species.

It is adaptive to know a lot about our partners, and it always has been. But there is now much more information and complexity to our relationships than at any other point in human history. Our ancestors didn't have to deal with a partner having an emotional affair via WhatsApp, or worry about whether or not their Hinge profile portrays their personality in a way that is most attractive to the algorithm. As the social and technological context of courtship and mating has changed, our behaviors have changed as well. Quite literally, we evolve with the times.

While some of the mating behaviors that occur in the natural world may seem strange to us (take the male argonaut octopus, who will tear off his own penis — technically that reproductive organ is termed a *hectocotylus* — and launch it at a female to avoid being eaten by her during mating), human reproduction and mate selection have evolved in response to similar biological forces and evolutionary pressures.[7] If, like the argonaut octopus, you're a species that doesn't form long-term pair bonds, every bout of intercourse comes with risk of predation, injury, disease, and even death — so there's more pressure for each mating event to result in offspring.[8] But for humans, the most impactful selective event is choosing the right partner — and this can take time.[9]

Due to our long gestation periods, and to the fact that human infants require several years of diligent, nearly around-the-clock care, human reproduction is a comparatively rare event compared to other species.[10] Historically, humans produced many more offspring across the course of a lifetime than people do today: Several centuries ago, it was not unusual for a woman to birth between eight and eleven offspring, according to some estimates. That number might seem high until you remember how high childhood mortality rates have been for most of recorded history.

In more recent years, humans are having fewer children, and, due to a wide range of factors including improvements in sanitation and healthcare and the development of vaccines against common and deadly childhood viruses, a larger proportion of those children are surviving to adulthood. Globally, women today have an average of 2.3 offspring per lifetime, and they are waiting later in life to have them.[11] Compare that to frogs, for example: In many frog species, males will fertilize anywhere from hundreds to thousands, even tens of thousands, of eggs as they emerge from a female, typically resulting in a high yield of viable spawn.[12]

Every animal needs a way to assess the reproductive fitness of a potential mate. Across ages, races, ethnicities, religions, genders, and sexual orientations, this decision is guided by evolutionary principles, even if we're not always aware of their influence. Disease avoidance, for instance, is a universal adaptation, something shared by people of all backgrounds.[13] If someone signals poor hygiene — for example, if they have bad breath or body odor — it's widely regarded as a turn-off. On a physiological level, we are being triggered by a disgust response that has evolved as a warning, or an evolutionary red flag, that someone may be physically or mentally ill.[14] Part of what's known as our behavioral immune system, this is our body's way of warning us to avoid someone who may pass their illness on, either genetically or through close contact, to offspring, or who may be unable to reproduce (get pregnant or impregnate) in the first place.[15]

Of course, in today's world, reproductive fitness is just one of many variables to consider when choosing a partner with whom we might want to procreate. Beyond whether we can have viable offspring together, we might ask ourselves, "Will this mate be parentally invested? Does this mate have the means to support me and my/our future offspring? Do they even want children, and if so, how many?"

These might seem like emotional considerations, but it is worth remembering a key fact: Emotions are themselves deeply evolved systems. They are necessary, purposive, and adaptive. And the architecture of our mind, our evolved inner animal, is shaped by adaptations that have long served our species. In other words, many of the traits associated with reproductive fitness or the capacity for parental care are ones we may find attractive and desirable in a partner, even if we don't intend to reproduce or co-parent with them. Moreover, research tells us that being with a partner with those desirable traits can change our attitudes toward wanting children.[16]

Even with our facility for sex without reproduction, modern humans must still approach mate selection with an eye toward the limits of our own resources. Before we get anywhere near deciding whether or not to form a long-term bond with another person, we must evaluate our emotional and financial capacities, our limited time, our emotional needs, our long-term goals, and our physical desires.

Because human reproduction is such a challenging and fraught process, our reproductive lives and mating behaviors are entirely bound up in long-term relationships.[17] We have evolved to invest in each other in order to survive. So, the story of human mating is also a story of human intimacy — which has all sorts of implications for our social worlds.

ALL IN THE FAMILY

For years I taught an undergraduate course on modern love. Whenever my students returned to campus after the first break of the semester, the first question I always asked was how many of their parents interrogated them about their romantic lives; invariably, lots of hands shot up. I assured them that such interactions don't stop even when you're into your thirties, as I could attest from my own experience.

There is no other species on this planet where kin, particularly parents and caregivers, exert as much influence over romantic and sexual choices as humans.[18] In fact, in one of our studies we found that close to one in four singles say it's important that your friends like your partner, and for a majority of young people between the ages of eighteen and twenty-nine, it's an absolute dealbreaker if your partner doesn't get along with your family. Anyone who's ever brought a new partner home to meet the family for the first time will know how significantly their approval (or disapproval, as the case may be) can impact the stability of a budding relationship.[19] Perhaps this is why daters in our studies also tell us that friends and family should respect their autonomy when their complicated reasons for choosing one suitor over another are at odds with the opinions of those others.

Much of the inherent tension in parent-child relationships centers on this conflict between wanting what we think is best for our loved ones and imposing that will on them, often at the expense of their own sense of autonomy.[20] And yet, evolution has programmed us to worry about and protect our offspring from the moment they are conceived, and well into their adult lives. The consequence of this type of sociality is that reliance on our families, and by extension our communities, is embedded both in our biological architecture and in the social customs that promote survival and reproduction.

Many of us take this social connection within which our reproductive lives unfold for granted. For instance, there is a popular misconception that breastfeeding is a naturally occurring behavior in humans, when in fact, we aren't a species for whom this most basic function of survival comes instinctively. Unlike other newborn mammals that can simply climb up their mother's fur and latch on in an open field, human infants have to be swaddled and held close to the chest when they nurse, and many women experience pain or infection during lactation.[21] In fact, one study that followed new mothers from birth through the first year of their child's life found that nearly 60 percent of those who stopped breastfeeding did so earlier than desired in part due to such complications.[22] For us, breastfeeding is a learned behavior, not an innate one.

This evolutionary paradox can be explained only by the fact that our species has long been enmeshed in cooperative breeding, a social system in which offspring receive care not only from their biological parents but also from alloparents, or "helpers," within their family networks.[23] Today, many Western hospitals will provide access to a lactation consultant or breastfeeding coach to help new mothers who struggle to get their babies to latch on. In other parts of the world, where family systems span multiple generations, young women learn, as they have for millions of years, from grandmothers, mothers, aunts, sisters, cousins. Family networks provide care to us offspring when we are young, but that nurturing impulse doesn't disappear when we grow up.

Kin influence can be explicit, as when friends or family offer an opinion (solicited or otherwise) that makes their dislike of a mate obvious, but some of its sway may be subconscious. We see evidence of this in what's known as similarity imprinting: our tendency to be drawn to partners who are in some ways similar to us and to our parents and caregivers.[24] One experimental study that used facial photographs of family members to demonstrate parental influence on mate preferences found a high degree of similarity between a husband's wife and his mother.[25] The study also showed that those who reported having rocky relationships with their parents were less likely to be attracted to someone with similar traits.

The closer we are to our parents, the more influence they exert on us. And the closer we are to our extended family, the more we care about their opinions on our potential partners. Notions of "extended family" can even extend to our pets, meaning Fido and Fluffy sometimes play a role in how we assess a mate's long-term viability. In fact, one study we conducted found that a significant proportion of both men and women (nearly half of the women and roughly a third of the men) had judged a mate based on his or her reactions toward their pet, and that straight women were more likely to be drawn to someone *because* he had a pet. Many people think of their pets as extended kin, and so a potential partner's ability to care for a pet demonstrates a capacity to care not just for themselves but also for others, and one day for us and perhaps our future children (human or furry).[26]

Our kin have always had a powerful influence on mate selection, but for modern humans who can open an app to choose from thousands of potential mates, it's a bit like being at a restaurant with one of those overwhelmingly extensive menus and turning to someone to ask, "What do you think? Should I stick with the salmon or try something new?"[27] When we ask people, "What do you think? Should I stick with the person I'm dating?," we might be asking for an honest opinion, we might be seeking validation for the choice we've already made, or we might simply be curious to know if someone else's assessment of our partner is the same as ours. In any case, the answer from a family member influences us more than one given by nearly anyone else, underscored as it is by the trust we usually grant our family and the protective impulse they feel for us. That doesn't mean that family assessments of our partners are necessarily accurate — our family might lack context, or overreact to minor issues out of a desire to protect — but they do shade our decisions in powerful ways.

That's one reason we might be reluctant to even mention a new partner to our parents or caregivers. We might not *want* their protection; we might fear their protective disapproval; we might not trust their judgment but know that it will have sway over us. That's why it is important, when we share news about a new partner with our family network or discuss our relationship, that we know what we're looking for and are clear about it: Are we asking for opinions? Are we looking for validation? Do we just need a sympathetic ear? Keeping in mind all of the protective and powerful dynamics at play within our kin influence networks can help us keep perspective.

I myself experienced the power of kin influence in my early twenties. I had just accepted a position that required a geographic move that would further my career but take me away from a relationship that was on the verge of getting serious. The woman I was seeing had initially been someone my family approved of. She was smart, ambitious, and well-educated, and she wanted to start a family and also have a career: a good match to me on paper in all kinds of ways. But over time, unbeknownst to me, my family had become increasingly unimpressed with the way she treated me. When it came time for me to decide whether to break up or try to maintain a long-distance relationship, I appealed to Jen, my trusted advisor, for advice. Although my cousin never explicitly told me what I "should" do, the more we talked, the more convinced I was that making a clean break was the right thing to do. It wasn't until after I had made my decision and was unpacking my new apartment that my mom voiced her relief, telling me, "I'm so glad you never had kids with that woman!" I knew at that moment that my family had talked about my relationship among themselves and that my mom's sentiment was broadly shared.

In retrospect, I realized that I had caught hints of their disapproval all along, and I have no doubt that on some level these stray remarks had affected my perception of the relationship and my behavior within it in ways that contributed to my decision to end the relationship. While I eventually was comfortable that I made the right choice, this experience provided firsthand evidence of how family opinions and influences matter to the bond between two partners. It's not *just* that the individuals make the couple and the couple makes the family; the family also helps the couple sink or swim.

BUT WHAT ABOUT SEX?

Because we are a social species that engages in intense pair bonding, sexual chemistry is not unlike adding yeast to a lump of dough. It takes more than that to bake a loaf of bread, but it's nonetheless a critical component of a recipe that produces a delicious slice — or, in the case of sex and intimacy, the alchemy of a deeply meaningful and enduring romantic bond. When we think about the role that sex plays in our relationships, particularly within an evolutionary framework, there's no escaping its purpose in reproduction — but we also now know it's doing a whole lot more than that.[28]

When Alfred Kinsey began compiling his first case histories in the late 1930s, sex was embedded within the marital construct. Americans in the first half of the twentieth century certainly knew sex was going on outside of marriage, but it was taken for granted that this was the place where sex *should* happen. When Kinsey approached his research on human sexuality, he and his colleagues recognized that a sexual event was different if it was with a spouse, a relative stranger, or a sex worker; they understood that relationship context matters, but they were focused on the taxonomy of sex and sexual behaviors, rather than the relationship context.[29] The *how* and *why* that context matters came decades later, and in many ways that is still an unfolding frontier in the science of human sexuality and relationships.

Today, the Institute he founded has expanded to include multiple labs across multiple disciplines.[30] We publish dozens of scientific articles every year — thousands in the past few decades — that have broadened the science of sex to include research on everything from the psychophysiology of sex to LGBTQ+ issues, polyamory, sexual violence, and more. We've learned a lot over the last three-quarters of a century, but there's still so much mythology around sex and sexuality, and still so many unknowns. For instance, we've learned a lot about orgasm — when it happens, why it's sometimes fickle — and yet there's still rigorous scientific debate about why we orgasm in the first place.[31] We know that there's a gender gap, but the evolutionary origin for this still eludes us. There have been dozens of papers debating the question of *why* human females orgasm when most of our mammalian cousins don't.[32] Is it a byproduct of our shared anatomy? A way to assess a partner? Does it have a reproductive benefit? On its face, it might seem like just a bunch of academics debating, but as we peel back the layers of this one question surrounding sex and gender differences in orgasm, we encounter a whole world of insight about the role of sexual behavior and sexual pleasure in human evolution — and what that means for modern relationships.

There's still so much we scientists don't understand about sex and sexuality, and yet it plays such a huge role in all of our lives — its presence, its absence, its quality, and the looming expectations and stigma surrounding it. What we do know is that relationship dynamics impact the types of sexual expressions we engage in, as well as the meaning, pleasure, and satisfaction associated with those expressions. Specifically, we know for sure that sex plays a big role in facilitating emotional bonds.

Humans have a capacity to engage in sex purely for pleasure. And we know that people around the globe, today and for centuries past, regularly engage in sexual activity that is not strictly focused on conception likelihood. We don't willingly engage in any other behavior that makes us more vulnerable, in every possible way. That's not a small thing. It tells us something about the evolutionary role of pair bonding. It tells us that at some point in our distant evolutionary past, there was some benefit to being able to engage in sexual activity with someone that didn't serve the purpose of reproduction.

When we're released from intercourse only serving reproduction, suddenly sex takes on another purpose, like drawing people closer together. Since humans — unlike most mammals — have sex facing each other so often (our most recent data suggests more than half of people engage in and enjoy this position regularly even if they're interested in trying other things), we can engage all of our sensory information. We can hear a partner's voice, inhale their scent, feel their breath, look into their eyes, and hold their face.

The fact that human anatomy allows us to do this (based on where our hands and faces and genitals are positioned) tells us that humans have evolved to engage in sex not just for the purpose of conception but also in the pursuit of intimacy, self-expansion, and even relationship maintenance.[33] Of course, sex is not the only path to intimacy; in fact, in our Singles in America study, we found that among the roughly two in five singles who have said "I love you" spontaneously, these words were more likely to be uttered over the phone (21 percent) than during (17 percent) or after (17 percent) sex. There are people who have little sex in their relationships and yet are very intimate, just as there are people who have a lot of transactional sex or casual sex without as much intimacy. But even in those cases, there are both sexual and nonsexual threads of intimacy that are difficult to totally disentangle, because sex and bonding are intertwined in our evolutionary history — and still present. As intimate animals, we are wired to catch feelings.

So, do we have sex to bond? Or do we bond so that we can have sex? Yes, to both. That these two processes are intermingled — psychologically, physiologically, and evolutionarily — is what makes the tensions between social and sexual monogamy so complex. We have these bonds with people we want to have sex with. For the most part we want exclusivity with them, or at least their focused attention, and they often want exclusivity with us. At the same time, we have that gnawing desire for sexual variety and exploration that evolution has left us with.

YOU BUT NOT ONLY YOU

Some years ago, I was invited to a houseboat party by a physician friend who thought I would enjoy meeting the folks who lived in a kind of floating community on a lake just a few miles south of Bloomington. He was right.

The boat slips off the little marina/bar-and-grill were crowded with friendly people. Boaters tend to be an amicable bunch, connected through their shared experience even if they've never met, smiling and waving to one another the way people who drive certain car brands honk as they pass each other on the road.

We headed to the last dock, where a cluster of large houseboats was harbored, many of which stayed docked most of the year. The margaritas were flowing, and people were laughing and grilling, their children hopping from boat to boat as the warm glow of the setting sun danced across the surface of the lake. I found myself delighting in this tight-knit community of entertaining people, whose lives and homes were so intertwined it was nearly impossible for an outsider like me to tell which houseboat belonged to which family.

Our host, Martha, was incredibly welcoming, and immediately forthcoming. We quickly slipped into that familiar pattern where someone realizes I'm "that Kinsey guy" and starts telling me personal things. Entirely unsolicited, Martha shared an unexpected sexual side effect of starting hormone replacement therapy while going through menopause. "I feel like I'm eighteen again," she laughed playfully; "sometimes the wind blows, and I get turned on."

I was happy to listen and learn. In many ways I'm still that five-year-old boy, magnifying glass in hand, observing fascinating creatures in the wild, trying to make sense of our complex social lives. As the night wore on, Martha's body language became more and more affectionate, more intimate — leaning closer, leaving a hand on my shoulder. At one point she even started to eat from my plate. I chalked it up to a party flirtation, though this didn't explain why her heavily muscled boyfriend kept smiling and winking at me, as if signaling I needn't worry. Toward the end of the evening, after Martha invited me to stay the night on their boat, my friend finally leaned over and whispered, "You know they're swingers, right? Most of them down here are. That's why I thought you would find them so interesting."

My friend wasn't wrong. Martha's invitation (which I declined, politely) highlighted one of the most fascinating paradoxes of human sexuality: We are wired to want variety. But at the same time, we also crave the intimacy that comes with exclusivity.

Is there any way we can eat our cake and have it, too?

One of the hottest topics in sex research right now concerns what's known as consensual nonmonogamy, a blanket term for relationship arrangements that don't require sexual exclusivity.

Consensual nonmonogamy has many varieties, from open relationships and polygamy to polyamory and ethical nonmonogamy (sometimes called negotiated nonmonogamy), each with its own specific nuances.[34] Polygamy refers to having more than one spouse, whereas a polyamorous arrangement involves a group of people who have multiple romantic bonds with each other, in contrast with an open relationship, where there is a primary relationship but multiple sex partners. There is also diversity within each type of relationship structure. For example, one couple might decide the relationship is open only when having threesomes together, whereas another couple might keep the primary relationship completely separate from other sexual partners. Yet another might be "monogamish," a relatively recent term for a relationship that is (as the name suggests) mostly monogamous but with flexible boundaries.

Though polyamory — the term, meaning "plural love," refers to a practice of having multiple romantic relationships — and open relationships are relatively new forms of consensual nonmonogamy, plural sex in other types of interpersonal contexts has also been well documented throughout human history.[35] Think about emperors having co-wives, or the practice of polygyny in various cultural and religious groups. In many cultures, plural marriage has been a long-established and accepted social practice, although in most cases it is gendered and only permissible for the husband to have multiple romantic partnerships. For polygamous Mormons in North America or among areas of West and Central Africa known as the "polygamy belt," for example, the wives do not typically take simultaneous husbands (polyandry) nor enter into romantic co-wife relationships with each other.[36]

Traditional women in the highlands of the Himalayas practice fraternal polyandry, where they marry sets of brothers, and in some societies, such as lowland South America and Amazonia, there is a practice known as partible paternity (or shared paternity), meaning that there is a cultural (if biologically unsupported) belief that multiple men have contributed to the conception of a single fetus.[37] Where partible paternity is part of the culture, women have sex with multiple partners.

Published research from our lab at the Kinsey Institute shows that about one in five single adults in the United States has had some kind of consensually nonmonogamous relationship at some point in their life, corroborating data hinted at in other studies.[38] However, when Americans are asked if they are *currently* in such a relationship, that number drops well below one in ten, which might suggest that more people *try* these arrangements than are able to sustain them long-term.[39]

Our studies found that while these numbers remain constant regardless of age, education level, income, religion, geographic region, political affiliation, and race, they do vary depending on gender and sexual orientation. Men were more likely to report having engaged in a consensually nonmonogamous relationship than women, and people who identified as gay, lesbian, or bisexual were more likely to have done so than those who identify as heterosexual (it's possible that LBGTQ+ people are more likely to experiment with new ways of ordering their romantic and sexual lives because they are already accustomed to navigating relationships that exist outside the dominant heterosexual norm).

Research suggests that interest in consensual nonmonogamy is growing. When feminist psychologist and sexuality scientist Amy Moors examined hundreds of thousands of anonymous search queries on Google between the years 2006 and 2015, she found searches for words related to polyamory and open relationships had significantly increased over time.[40] What this suggests is that the topic has entered the public's consciousness and generated questions for many people. This does not tell us whether the number of people actually *engaging* in consensual nonmonogamy has increased over time, which would be a rather difficult question to answer given existing data records.

When people ask me about polyamory or consensual nonmonogamy today, I usually sense that what they're *really* asking me for is permission. They want to know if it's okay.

BEYOND THE PAIR BOND

Though I am of the firm belief that there is no right or wrong when it comes to consensually negotiated relationship dynamics, therapists have traditionally cautioned against pursuing consensual non-monogamy because those relationships were viewed as unstable or unsustainable. However, the newest body of scientific literature argues that people in more sexually "open" relationships don't fare worse psychologically or emotionally than people in monogamous ones.[41] And a series of studies have deduced that while consensually open relationships might not work for everyone, or even for most people, there are many people for whom they do work perfectly well.

Evolutionarily, though, these exceptions might prove the rule. For some, consensual nonmonogamy is very fulfilling, and it helps to relieve some of the tensions between love and sex that live in our intimate pair bonds. But it also introduces new tensions, and I've come to believe that the strain of those tensions is more than most people can or want to handle. That strain may result from cultural stigma, or from the general neurobiology of pair bonding, or from the fact that intimate love is characterized by focused attention on the object of our affection. In general our brains don't appear particularly well suited to processing intimacy with more than one partner at a time.

Despite the research that identifies threesomes as the top sexual fantasy (interestingly, for many, this fantasy involves their primary partner along with one other), actually taking the steps to make such a fantasy a reality is a challenge for many.[42] Even casual polyamorous encounters take substantial effort and negotiation, and communication is key to maintaining the bonds that any intimate experience requires. Who needs more touch? Less? Who is feeling neglected? Who needs more time with whom? What is the state of things between each member of the polycule and each of the others? Often, these variables put even more pressure on the intimacy bonds that we form with our primary partners.

One of the prevailing rationales for consensual nonmonogamy is that some among us simply have too much love to give to just one person, but the reality is the opposite: Most people don't have the biological, psychological, and social tools to love more than one person at a time. Consensually nonmonogamous relationships require considerable effort, and basic demographic data reveals they do not appear to work for most people today. This is likely because these arrangements are in many ways incongruous with our evolved tendencies for pair bonding, as well as our social-cultural systems around intimate relationships. In short, the intimate pair bonds we have exhibited throughout our history (the ones that we have evolved with, and that have allowed us to thrive as a species) are difficult to sustain with multiple partners, especially over the long run.

This general pattern does not, and should not, invalidate those who find extraordinary happiness in their nonmonogamous relationships. There are all sorts of reasons that opening a relationship sexually can really help preserve a love bond — someone is deployed for military assignment, someone has a severe sexual dysfunction, someone travels for business, someone has a medical condition, and so on. In the short term, these arrangements can be a way of "having it all," of expanding one's sex life without expanding (or challenging) one's romantic partnership, allowing one or both partners to enjoy more sexual variety and intimacy without having to invest a lot more emotionally. But I suspect that, over time, the intimacy most people experience in consensual nonmonogamy is more sexual than it is emotional, and the same issues that plague monogamous relationships — mismatched libidos, jealousy, boredom, and more — tend to surface in consensually nonmonogamous ones. In fact, they can *multiply,* given the effort of maintaining intimacy across multiple partnerships. So while some people find ways to open their relationships and receive much joy, that is not the most common pattern, and in my view it is not a realistic option for most people. I would even venture to say that for the majority of people, it is simply not possible to experience sustained *love* across multiple partnerships, or at least not for long.

As one of my friends who had attempted to form a polycule once told me, "It didn't work. I just pissed off two women instead of one."

Which brings us back to where we started. If we can't maintain love and intimacy within a monogamous relationship, we might not be set up for success in a consensually nonmonogamous one.

LOVE ON THE BRAIN

My friends Justine and Patrick have maintained a passionate relationship for more than two decades. After all these years, Justine still laughs so hard at Patrick's jokes that people dining three tables over will turn around and stare. It's the kind of deep, joyful belly laugh we should all be so lucky to have twenty years into a relationship.

What's their secret?

One of the hallmarks of romantic love is a tendency to view each other through rose-colored glasses. So Justine may think Patrick is hilarious even though not everyone is as entertained by his food puns ("You wanna pizza me?") as she is. This tendency to view our partners more positively than others might is what behavioral scientists refer to as "positive illusions," and the more of these positive impressions people hold about each other, the stronger their bonds tend to be. In one meta-analysis of 137 different studies conducted over thirty-three years (with a total of 37,761 participants), the more positive illusions individuals reported experiencing toward their relationship partner, the less likely they were to end their relationship.[43]

In other words, narratives about how extraordinary our partner is can help keep the love alive; at the same time, the reason we create these narratives about our partner in the first place is precisely that we love them.[44] Positive illusions are a mechanism for how couples begin that process of psychological overlap we see measured in the Inclusion of the Other scale mentioned earlier, and for couples like Justine and Patrick, it's also key to maintaining intimacy even through all the ups and downs.[45]

The ability to create narratives — particularly positive narratives — about our relationships is evolutionarily adaptive. When we can tell ourselves a pleasing story about our relationship, we are more easily able to invest in our partner and focus on improving our relationship with them. In one fMRI brain scan study, seventeen married people were recruited, all of whom had been married for at least ten years and responded positively to the question "Are you still madly in love with your long-term partner?" Among these participants who reported *long-term* intense romantic love, researchers found that for those who maintain feelings of passionate love, a variety of brain regions are involved, including the reward systems associated with dopamine and early-stage romantic love (the ventral tegmental area and dorsal striatum, for the technically minded) but also neural systems associated with attachment and long-term bonding (the globus pallidus, substantia nigra, thalamus, and insular cortex).[46] We know from earlier research that early-stage romantic love is associated with activation of the first set of neural systems and brain regions and that more affiliative love, such as maternal bonds and friendships, is associated with activation of the second set. So, for those who experience long-term intense romantic love, there appears to be a melding of these feelings and also of the biological responses at play in the brain and body.

The activation of particular brain regions in turn impacts our nervous system's reactions to our social world. When psychologist Jon Maner and colleagues looked at how people in committed romantic relationships respond to seeing attractive people of their preferred gender who were not their partner, they found that after being asked to think about their romantic partner, people would avert their eyes and gaze from attractive others, without awareness of what they were doing.[47] This is how our body regulates the tension between our sex drive and our evolutionary propensity to pair-bond, reinforcing monogamy by directing our focus back to our mate. Doing so is adaptive; it helps ensure that we nurture the bonds that help our species survive.

Chapter Six

NEST

Jen waited until after their wedding day to move in with Dave. Once I was old enough to notice, that decision seemed unusual, like a leap of faith or buying a car you haven't test-driven. But it turned out that whatever quirks they might have discovered about each other from living together weren't enough to shake the bond they'd already formed. Many years later, I would come to see that their decision to wait until marriage was an expression of their faith in that bond. That, of course, isn't the case for all couples.

For millions of people all over the world, "living together" has a variety of implications for their relationships, from sharing the same bed and synchronizing sleep patterns and circadian rhythms to learning more about each other's habits and quirks.[1] There are also economic factors; for many couples, the money saved on paying individual rents can be put right back into supporting their goals and passions, as well as just generally doing fun stuff together.

It once was widely the case that you lived with someone only if you were married, or at least engaged to be married. In the United States as recently as 1968, the idea of out-of-wedlock cohabitation was still scandalous enough that one young woman was expelled from Barnard College in New York City after it was revealed she was living with her boyfriend.[2] Cohabitating outside of marriage initially rose in popularity throughout Western Europe, especially Scandinavian countries, during the latter half of the twentieth century, and the pattern has spread across the globe. By the time Jen and Dave moved in together in the late 1990s, their choice to wait until after marriage to do that already seemed a little old-fashioned, and it's a relationship decision that's becoming less and less popular every year.[3]

In the United Kingdom, nearly 4 million unmarried couples choose to live together; in Australia, nearly four out of five couples have lived together before marriage; in China, more than half of millennials report cohabiting with their partners prior to marriage (a nearly fourfold increase from two decades ago); and in the United States, the percentage of adults who have ever lived with a partner outside of marriage (59 percent) has now surpassed the percentage of adults who have ever been married (50 percent).[4] According to the Pew Research Center, this is a trend being set by young people. Their analysis of 2017 U.S. data found that just 4 percent of American adults fifty years and older were cohabiting, compared to 12 percent of those younger than age thirty. And according to a Pew Research Center analysis of data from the 2013–2017 National Survey of Family Growth, those aged eighteen to twenty-nine are nearly twice as likely to have cohabited (44 percent) as they are to have married (23 percent).[5]

Other couples who live together fall into the category of domestic partnership or common-law marriage, which in some jurisdictions provides couples with legal rights and privileges after a certain number of years. These marriage-like arrangements were originally more common in same-gender (gay and lesbian) relationships prior to the enacting of marriage-equality legislation, as happened in the United States in 2015 when the Supreme Court struck down state bans on same-sex marriage, legalizing it in all fifty states.[6] Protections like these have opened the door for heterosexual and same-gender couples to explore alternative pathways of codifying their romantic relationships in the eyes of complex legal systems worldwide.

In most laboratory animal studies, it's important (and often required) to house social animals with others, because isolation can be cruel and damaging to health and behavior.[7] For social animals like humans, living with an intimate partner can provide many of the social connections we not only crave but also need to survive.

BUILDING A NEST

On quiet weekends, Michelle and I will take the day and go for a hike through the woods and rolling hills of Brown County State Park. Everywhere we look, we see signs of nests. A patch of tall grass where a doe and her fawns might settle for the night. A clump of twigs and moss in the hollow of a tree housing a family of squirrels. Even the elderly couple holding hands as they make their way along the meandering trail seem to all the world like they are entirely at home.

All nests are about animals coming together, whether for a few moments, a breeding season, or in some cases a lifetime. To humans, a "nest" is more than a physical home. The word itself inspires all manner of metaphors. When expecting parents spend time setting up a crib, decorating a nursery, and generally preparing for the birth of a newborn, they are "nesting." When a teenager goes off to college or is otherwise encouraged by their parents to move out of the family home, they are "leaving the nest" and said parents become "empty-nesters." Symbolically, this is not so far off from the way some species of seals will nurse their young for a few weeks, then shove them off the ice floe so that they learn to swim — and survive (as a metaphor, "shoved off the ice floe" did not catch on).

Some animals are solitary and don't "do" nests, but species that pair-bond tend to display nesting behaviors not unlike the ways in which we make our own homes with each other. One way to think about what is happening, on a fundamental biological level, when humans decide to build a nest together is through the lens of life history theory, a framework that marks the major evolutionarily relevant milestones an animal experiences across its life course.[8] In humans, birth, moving from lactation to food, puberty, leaving a nest, age at first sexual activity, and age at first reproduction are all important markers that significantly impact the process of development and growth.

For millions of years, women spent the vast majority of their lives moving between nursing and reproduction. According to some models, in many preindustrial societies the average woman had, statistically speaking, between five and nine children over the course of her life.[9] This is a revealing point that brings into focus how modern contraceptives have liberated women from the certitude of childbirth and redefined the very notion of when and how we enter into pair bonds. But researchers have also highlighted that population fertility patterns vary considerably, including for our preindustrial ancestors, based on particular environments, resource and food availability, relationship dynamics, cognitive decisions by individuals and couples, family pressures, and health and disease.[10] One of the highest recorded fertility rates was among early twentieth-century North American Hutterites, where married women had an average of approximately eleven children. In modern times, according to the 2024 *World Factbook,* Niger has the greatest fertility rate, with an average of 6.64 births per woman, and the parts of the world with the lowest are Taiwan and South Korea, both with an average of 1.1 births per woman.[11]

As the work of evolutionary demographer Rebecca Sear has shown, our ancestral grandmothers usually had many more children than most contemporary populations, starting in their late teens and spanning their adult lives; the interbirth intervals provided just enough space for the pair-bonded parents to devote the resources of the nest to the survival and optimal care of each of their young.[12] Compared to our primate cousins, and especially compared to those with mating systems that lack paternal care and cooperative breeding, human interbirth intervals are relatively short. While short interbirth intervals may allow individuals to have more offspring, having a very large number of offspring comes at a cost in terms of the quality of care, attention, resources, and energy that can be devoted to each offspring. Thus, the adaptive strategy of our species, in terms of promoting survival and longevity, is characterized by enough distance between successive offspring, with continuous dual parental and familial investment.

Having what's known as altricial young—offspring that require a lot of caregiving—makes pair bonding a necessity: We need help to raise children.[13] Pregnancy, infant care, and parenting are time-intensive practices for humans. Forging long-term romantic bonds allowed our forebearers to have offspring that required greater investment for longer periods of time, in turn allowing for the evolution of more sophisticated human brains and cognition.

We also need a stable, safe environment free from threats. Humans evolved to nest, in part to protect young. But our impulse to create a safe and nurturing nest is also beneficial to the pair bond itself, with or without children. When we nest, we create a safe space where we can support and protect each other.

Today, these "nests" come in many shapes and sizes, and our orientation to them remains central to human survival.

IT'S A FAMILY AFFAIR

I was raised by my single mom until I was thirteen, when she married my stepfather, Mike — whom I've long called Dad. There were more than fifty apartments in our building, and one thing I learned from her was how to live in close proximity to other humans without too often lowering the drawbridge to our personal space. Growing up, we didn't really know most of our neighbors except to say hello in the elevator or by the mailboxes.

One of my mother's neighbors, who has lived a floor away for over thirty years, has never stepped foot inside my parents' apartment. "If you let them in once," Mom used to tell me with a mostly serious smirk, "they'll think it's okay to come by all the time."

Throughout evolutionary history, mothers have played a central role in building, maintaining, and protecting the family nest. The ancestress hypothesis, an evolutionary theory proposed by anthropologist Kathryn Coe, describes a cultural mechanism by which information was passed down through the mother-daughter line via visual art forms, like body decoration and ancestral paintings, in order to strengthen and reinforce kinship bonds.[14] This theory may also help to explain why mother-daughter dynamics are so often fraught: Evolutionarily, they are forged to be intense in order to enable this matrilineal transfer of important information.

Of course, mothers aren't the only influence on the family nest. For millions of years, humans lived in small communities or tribes. Within those groups, people formed families in which pair-bonded individuals lived with their own children, but also with parents and grandparents, aunts, uncles, and cousins, who provided a source of shared support and resources. Like the northwestern crows who nest in family groups, with older offspring helping their parents to raise each season's new brood, we too have always been a cooperative breeding species.[15] It was necessary if children were to survive in a world where competition for resources was fierce. And those of our ancestors who survived were the ones who invested in family units.

In many societies, these units might also include nonbinary or gender-fluid family members. The indigenous Polynesian people of the Samoan Islands, for example, recognize men, women, and a third gender, the *fa'afafine*—natal males who practice femininity and typically remain in the home they're brought up in to care for their parents and extended family.[16] Studies of this population by behavioral scientist Paul Vasey and others found that families where a *fa'afafine* (which roughly translates to "in the manner of a woman") inhabited the role of alloparent—a caregiver who is not a natal parent—for their nieces and nephews do better in terms of survival and reproductive outcomes.[17] In other words, it is evolutionarily advantageous at the level of a family to have this avuncular investment. Even if the *fa'afafine* aren't reproducing themselves, they are an additional caretaker contributing to the family's genetic legacy.

A major gap in the literature, and one my research team and I are working to better understand, is how these factors are different in same-gender-parent households compared to different-gender-parent ones. Older studies showing that children raised by same-gender parents have worse health and well-being outcomes have largely been overturned, due to methodological limitations or statistical irregularities with the original studies. In fact, more recent research has shown that children raised by different-gender parents and same-gender parents have similar outcomes. In some studies, we've even seen that children raised in same-gender-parent households (with two dads or two moms) actually perform better in primary and secondary education, which researchers hypothesize might be because these parents feel the need to overinvest in order to mitigate any stigma or backlash from those who might expect (or, sad to say, want) to see them struggle.[18]

Recent research has also challenged views that monogamous households necessarily lead to better outcomes for children and families. In Western countries, children raised in more diverse family structures have both unique challenges and opportunities — on one hand, they face social stigma, but on the other, they reap the benefits of additional caretakers.[19] In one study utilizing twenty years' worth of anthropological findings among the Mpimbwe people of rural Tanzania, where both polygyny and serial monogamy are common, monogamous marriages were not regularly associated with better developmental or educational outcomes for children.[20] This highlights the fact that the human animal can adapt to many different types of social structures.

In the twenty-first century, we are reconfiguring the notion of the nuclear family.[21] Indeed, a major change has come in the way people today tend to structure their living arrangements as new generations push for greater independence.

Many millennials, in particular, have forged a new path, living by themselves or with friends instead of in family units well into their twenties and even thirties. Young people might stay in the same neighborhood, city, or state as their families, but not the same house, whereas in the not-so-distant past children who moved out tended to move away, and usually only once they'd found a partner. Which on average they found at younger ages than today.

At the same time, we've also witnessed the rise of what's become known as the "boomerang generation," a designation that refers to young adults who return to their childhood homes after graduating from college, a trend that appears to be largely driven by the difficult financial realities faced by so many young people today.[22] This reverse migration was especially prevalent during the pandemic, when nearly a third (32 percent) of millennials and Gen Zers moved back home with their parents.[23] Slightly more than half of those who moved home say it was out of necessity, with the other 49 percent saying it was by choice — that coming home at a time of such profound global uncertainty just felt safer.[24] As recently as 2024, data from the U.S. Census Bureau show that roughly one in three people between the ages of eighteen and thirty-four lives in their parents' home; according to a Bank of America survey that same year, nearly half of young adults said they didn't earn enough money to live the way they wanted to on their own.[25]

In a funny way, this is an evolutionary echo of the ancestral past, when it was advantageous to live in multigenerational homes that included parents, grandparents, siblings, and children, all of whom contributed to the functioning of the household and to one another's social and emotional needs.[26]

When we peel the metaphorical onion of the family down to the center, however, we find the dyad. There are exceptions to the rule, but even in poly relationships (or other consensually nonmonogamous arrangements), where there might be multiple dyads and arrangements, groups of two predominate. And even most single-parent families originated with some form of the dyad. As the pair bond solidifies, our bodies and brains have evolved to physiologically and psychologically shift from "I like you, let's spend more time together" to "I love you, let's build a nest together." Making this leap is a significant milestone in the relationship, ushering us into a new phase of our intimate lives.

KEEPING THE FLAME ALIVE

Even in strongly pair-bonded couples, sexual passion wanes over time. Researchers have found that somewhere around eighteen to thirty-six months into a relationship, what some might call a "cooling down" tends to happen. On average, we move from intense feelings of attraction and being in the thrall of obsessive, intrusive thoughts into a more attached and comfortable stage of the relationship. What we think of as passionate love, in other words, begins to transition to companionate love. Think of this as the "couch-and-cuddle" phase.[27]

If passionate love is a hormonal hurricane, companionate love is the calm after the storm. One stage is not necessarily better or worse than the other. Some couples find themselves growing more distant once the flames of sexual passion begin to subside. Others look forward from the start to transitioning from those turbulent initial stages to couch-and-cuddle.

Nor is it necessarily a linear journey. It's simply the warp and weft in the fabric of human intimacy. Once the companionate-love phase sets in, the emotional bonds in the relationship often grow stronger; as pioneering research by social psychologist Elaine Hatfield has shown, for most people, companionate relationships feel safer, more stable, and more comfortable than other types of love.[28] Yet for many long-term couples, entering the companionate phase marks a decline in sexual frequency, as the intensity calms and partners locate other ways of connecting, often more deeply and intimately. So the question before us becomes: Is it possible to sustain sexual passion throughout the life of a relationship?

Given all the complex and sometimes contradictory evidence in behavioral research, ethnographic accounts, and the clinical literature, my colleagues David Fredrick, Janet Lever, Brian Gillespie, and I decided to explore this very question. In a landmark study of almost 40,000 heterosexually coupled people who were married or cohabiting and who had been in their relationship at least three years, men and women of all backgrounds responded to a detailed questionnaire on their habits and feelings of passion and sexual satisfaction.[29]

When we focused in on those men and women in our study who were more likely to report that their most recent sexual encounter with their partner was "passionate," "loving and tender," or "playful," we found that they engaged in a suite of specific behaviors largely associated with intimacy and closeness, including giving or receiving a massage, going on a romantic getaway, or planning a date night to have sex. Moreover, about half of the more satisfied participants, of both genders, reported their last sexual encounter had lasted more than thirty minutes, which likely reflects their reports of more time dedicated in general to mood setting, foreplay, and behaviors beyond intercourse alone.

At the same time, we found that those partnered individuals who have kept the embers of sexual satisfaction smoldering over the years also reported behaviors like trying a new sexual position, talking about or acting out fantasies, trying anal stimulation, or using a sex toy together.

On the surface, this might read as though the trick to sustained passion is simply lots of sexual variety, but I believe these findings actually tell a story of intimacy. Wearing sexy lingerie, setting the mood with candles, taking a shower or bath together, or experimenting outside of your usual comfort zone are all ways of demonstrating interest in being intimate with a partner and in so doing communicating trust, vulnerability, and connection. It becomes more than just any one particular act of sex. The power is in the commitment of effort in advancing passion, together.

We had known going into the study that frequency of sexual activity tends to decline as a relationship goes on, but we found that even though people in long-term relationships might have sex less often as the years go by, their sexual satisfaction can actually improve over time. In other words, our work demonstrated that while most people felt that their passion declined over time, it is also true that if passion is nurtured in a relationship, it can stay alive for decades.

Given the extraordinary amount of data we were able to pull from the original study, which examined only heterosexual-identifying men and women, we launched two more studies. One asked the same questions of women in relationships with other women, and another was directed toward men in relationships with men. Using an advanced statistical analysis technique called coarsened exact matching, we compared how satisfied heterosexual women were compared to lesbian women and how their satisfaction levels mapped onto experiences with over seventy-five sexual practices. And we used a similar list of seventy-five practices to gauge and compare the responses of heterosexual men and gay men.

In the partnered-women study, lesbian and straight women were equally likely to be sexually satisfied, but lesbian women were more than twice as likely to report having sex never or infrequently; this pattern was magnified in relationships longer than five years (42 percent of lesbian women versus 15 percent of heterosexual women). This drop in sexual frequency has been colloquially popularized as "lesbian bed death" — a term criticized by many sexologists — but when we dug deeper into this large dataset, we found that while these couples were having less frequent sex, the sex they were having was more intimate: Lesbian women were more likely to orgasm regularly, receive oral sex, and use sex toys. In their most recent sexual encounter, compared to straight women, lesbian women with higher rates of orgasm were also more likely to say "I love you," have sex for longer than thirty minutes, and engage in gentle kissing. These intimacies likely explain why sexual satisfaction rates were similar in these groups of women despite notable differences in sexual frequency.[30]

Among gay men, we found similar patterns of intimacy, but without the sharp decline in the frequency of sex. Gay men were more likely than straight men to receive or perform oral sex regularly, to have had anal sex during their most recent encounter, to have viewed pornography with their partner, or to have invited another person into bed with them and their partner. Gay men were also less likely to report a sexual-desire discrepancy — that is, a scenario in which they wanted sex more (or less) often than a partner — than straight men, particularly in relationships longer than five years.[31]

Though the variety and frequency of sexual activities may have varied across the three groups that we studied, all had one important thing in common: Communication, both in the bedroom and in the relationship in general, was a strong predictor of sexual satisfaction. That pattern held, regardless of gender or sexual orientation. In other words, those couples who were keeping the flame of intimacy alive by employing a greater number of strategies (texting, talking, leaving notes, body language) to make sure their partner understood what they were thinking or feeling were keeping the flame alive sexually as well.

GOING NUCLEAR

Evolution has primed our species to follow a pattern: that we live and make a nest together in order to survive and reproduce. Because this information is passed down to us from our own family, many people end up building a nest similar in structure to the one they grew up in. But this is not a pattern everyone wants or follows. As always with our natural history, there are exceptions to the pattern, and some consciously choose another path.

Now more than at any point in history, as humans have gained the ability to control our own reproductive destinies, we are better able to consider economic issues that inevitably impact our ability to provide sufficient resources to our utterly dependent offspring. We ask ourselves: Is our home big enough to add another member to our family? Can we afford to send two kids to college? Or, more germanely, do we *want* to take on the responsibilities of raising a family for the next thirty or forty years? That many people worldwide are answering no to these questions explains why we are seeing much lower fertility rates than we have in the past. After accounting for infant and child mortality rates, an average couple today would need to produce more than two children to maintain replacement fertility, a demographic term that refers to the level of fertility at which a population replaces itself from one generation to the next. Yet, almost half of the population of Earth lives in a country where people are not reproducing quickly enough to actually replenish the existing number of humans in their own community.[32]

Research by historians and anthropologists, including my colleague Rebecca Sear, who directs the Centre for Culture and Evolution at Brunel University of London, highlights that many people today have a particular perception of the structure of a traditional nuclear family.[33] But here's the catch: That structure is a myth. The notion of a family nest with 2.5 children, a male breadwinner, and a woman as the sole domestic laborer is not a prototypical model when we look at societies and families around the world. In fact, such a model is no more likely to produce familial and relationship happiness than the myriad permutations of "nesting" arrangements we see across the globe today.

Humans are adaptable to their ecology, and so too are the structures we use to define and build our nests. It's that very flexibility that is part of the human story of adaptive variation.

And at the center of this variation is the pair bond.

As compared to singles, married people in our studies report lower amounts of anxiety and lower amounts of stress across a variety of "life domains" — a term psychologists use to describe the aspects of life that are central to the human experience over the course of a lifetime.[34] The last decade in particular has witnessed a growing amount of research specifically addressing whether the benefits associated with marriage stem, in fact, from the marital union — or from the act of cohabiting.[35] Put another way, are these marital benefits a response to the intimacy that arises from sharing a nest with another person, or a response to the cultural practice of being married?

On this question, researchers have duked it out for decades. Some sociologists studying the social and economic pressures to marry — along with the stigma attached to singles and singlehood — have argued that many of the studies on the benefits of marriage are deeply methodologically flawed (the ongoing intensity of this debate is an important reminder that we should be cautious about how we interpret the data in our pursuit of an understanding of what it all means in our daily lives).[36]

For decades, researchers have similarly debated links between cohabitation and divorce, generally showing that premarital cohabitation is associated with higher risk of divorce and lower marital quality. But in recent years, some North American studies questioned this effect, arguing that cohabitation had become so common that these patterns appeared to be diminishing, if they had not yet dissolved altogether.[37] Recent research suggests that once all other factors are controlled for, couples who cohabitate premaritally are less likely to divorce in the first year of marriage, but beyond the first year there is still a moderately increased risk of divorce.[38] In other words, while we might think cohabiting provides the body and mind with all the benefits of getting to know and connect with a partner, a more accurate reading might be that cohabiting provides physical and psychological benefits that lie somewhere between what we observe in singles and what we see in married people. While premarital cohabitation has advantages, there also appears to be something about the social and cultural practice of marriage that results in higher relationship satisfaction and better health.

Of course, the two statuses, premarital cohabitation and marriage, have certainly gotten murkier over time.[39] While there may be more variability in the commitment of the couples involved, cohabitation affords advantages in the form of opportunities for sexual activity and emotional closeness. These behaviors can come together to form something of a special sauce for human intimacy — one that becomes even more crucial if and when children arrive.

SEXUAL SATISFACTION AND PARENTING

It's widely believed that relationship and sexual satisfaction declines sharply during pregnancy, around childbirth, and during a new baby's early infancy. And yet, in one Canadian study of 203 couples, nearly half retained high commitment and high satisfaction, meaning that pregnancy and childbirth do not inevitably mean trouble for all, or even most, couples.[40]

Many couples have less sex after having children, and this is for both biological and emotional reasons.[41] During pregnancy, some couples report concerns about sexual activity potentially impacting the health of the pregnancy, leading to lower sexual frequency. For others, behavioral and relationship changes during pregnancy and postpartum are associated with changes in sexual desire, often decreases. Following birth, some new moms also experience lactational aggression, also termed maternal aggression, consisting of adaptive temporary defensive behaviors to protect offspring. Evolutionarily, there is little incentive to reproduce again immediately after giving birth — doing so would reduce a mother's ability to care for and nurture the new offspring.[42] Additionally, in the postpartum period women are (obviously) exhausted, overwhelmed, and sometimes even physically injured after childbirth, with their bodies and brains needing to recover from an immensely taxing physical experience.[43] While there are technically no clinical requirements for how long a mother should wait after childbirth to resume sexual activity, most healthcare providers recommend four to six weeks to allow bodily healing and reduce the risk of complications, which is highest during those first few weeks after delivery. Parenthood has all kinds of effects on our romantic and sexual lives, everything from the impact of sleep disruptions to the physiological trade-offs of nursing, during which women's sexual motivation tends to decrease.

And men are certainly not immune to sexual and hormonal changes during this transition. Drops in testosterone associated with fatherhood have been demonstrated in a range of populations.[44] In a series of illuminating studies by my collaborator Peter Gray, he and his team have shown, in samples of men from across the globe, that testosterone has been a particularly important variable in influencing the physiology and behavior of "mating effort" versus "family effort."[45] In fact, in studies of new fathers and in others where men are asked to hold and swaddle babies, we see decreases in testosterone levels (and increases in oxytocin).[46] Why? Scientists theorize it's because these dads are in family mode and don't need elevated levels of androgens — the sex hormones, including testosterone, that play a role in kick-starting puberty and reproduction, and that prepare them to be ready to fight or mate. Their bodies are mobilizing a different physiological response in the presence of the child, one that motivates them to nurture and care for their offspring.[47]

One finding from our Singles in America study that seems surprising on the face of it is that among single parents, those who had children under the age of five years went on more dates and had more sex than those with kids ages five to eighteen.[48] At first, we wondered why this would be. Finding childcare for an infant or toddler so you can go on a date is often a significant challenge for single parents. But as we thought more about our findings, we realized there is likely an evolutionary reason for this pattern. We hypothesized that singles with very young children are more motivated to find a partner with whom to co-parent (and nest) than those whose children are more self-sufficient.[49]

For women, the gender disparities that emerge once children have entered the picture can lead to resentment and, in turn, lower relationship and sexual satisfaction. Many mothers do double duty, working during the day — whether that's a job that brings in an income or the domestic work and child-rearing that don't — and then cooking and taking care of the rest of the household in the evening. Today we call it the "second shift" of domestic and emotional labor, and we use the term *labor* very explicitly to highlight that this is *work*.[50] According to the *New York Times,* between August and September 2020 more than 800,000 women in the United States left the workforce, presumably to meet the demands of childcare and virtual learning during the height of the COVID lockdowns.[51]

Meanwhile, a survey taken earlier that year found that almost half of the fathers queried thought *they* were doing most of the homeschooling — but only 3 percent of mothers agreed.[52] One interpretation of this discrepancy might be that the men were overvaluing their own contributions to the household and undervaluing their partner's, a residual effect of the greater gender disparities that are still deeply embedded in modern relationships.

But household disputes over the division of labor are far from the only source of stress or tension a couple may have to navigate once they build a nest together. And having a strong foundation of intimacy becomes more important than ever when we fall on difficult times.

IN GOOD TIMES AND BAD

Jen and Dave built their first nest in Dave's beautiful one-bedroom apartment in Forest Hills, not far from the apartment building where I grew up. It was a posh, well-maintained doorman building, a space they owned together. A few months later the cockroaches appeared. Jen freaked out.

"We are *not* living here," Dave remembers her saying. Jen was twenty-four years old, and the roach situation was the first time she put her marital foot down. The timing, it turned out, was actually pretty good. It just so happened that Jen's mother, Mary Jane (my mother's only sibling), had a vacancy in the upstairs apartment of the two-family home in Atlantic Beach where Jen grew up. Without further ado, the newlyweds said goodbye to city life and city roaches and headed to Long Island.

The white stucco house was set back from the road by a short driveway — at the end of which I had many years prior conducted my early (un)scientific studies of sidewalk ants. It had a long garden, lined with tall grasses and strawberry plants in the summers. An ideal nest in which to raise a family. When I visited as a child, Jen and I would pick those strawberries, then climb the big flight of gray wooden steps to the front door, head to the kitchen sink, and eventually sit on the rattan furniture in the living room and eat what we'd harvested.

As pediatric health expert Thomas Boyce has described, there are children who are "orchids" and require careful tending to realize their potential, and there are "dandelions" who can thrive in just about any conditions.[53] Jen was the ultimate dandelion. This was true when she was growing up in a family that moved around a lot, it was true when she settled down with Dave on Long Island — and it was also true when she was diagnosed with brain cancer shortly after the birth of her and Dave's second child and their family nest was turned completely upside down.

The natural progression of a romantic relationship, after all the searching and the choosing is done, is to build a nest with someone. The thing that no one tells you is that the nest, this consequence of the evolution of a relationship, can quickly become the greatest threat to the pair bond. Nesting fundamentally changes a relationship. As partners face new challenges, the pressure points can be structural, like navigating the responsibilities of parenting and shared domestic labor, or unexpected, as when we come face-to-face with a tragic accident or illness.

These pressures can, in fact, blow up a nest, especially as we struggle to maintain and sustain our emotional bonds in the face of our often-diverging impulses for romantic commitment and sexual variety.

Chapter Seven

STRAY

As Sheriff Walter Harrison stepped from the garage into the kitchen, his lawman's gut tingled. Something was off. Wasn't the kitchen light normally on when he got home? Had there been an unfamiliar car parked out on the street? Following his instincts, he went to the bedroom. The door was shut. Feeling as if he were sleepwalking, he opened it.

It had been a quiet day for the sheriff. In his sleepy Illinois township, the most common offenses were drunk driving, unlawful hunting, and teenage mischief, and on this particular day there hadn't even been any of that to police. So he had peeled off early, wishing the officers in the station a good afternoon. He had worked hard for two decades to become the boss; he could allow himself this small luxury every once in a while.

Plus, lately he'd been feeling there was some disconnect between him and Janice, the hazel-eyed, sharp-as-a-tack woman he had married sixteen years earlier. Their schedules weren't as compatible as he would have liked, and even when he was off duty, he often got called into the station. Today, he thought, he would surprise her, and they could go out for dinner at Spagnuolo's, their favorite Italian place, which they hadn't visited in ages.

When he opened the bedroom door, however, time stood still.

There was Janice. In bed — *their* bed — with a man. Sheriff

Harrison looked at his wife, who made a sound that was the beginning of a word. She had the sheet pulled up to her chin, and she looked at him in that moment with a mixture of shock, guilt, and defiance. Then something inside him just...snapped. Still wearing his uniform — the brown collared shirt with a brass star pinned above the left breast pocket — he pulled his standard-issue sidearm out from the holster at his hip, aimed, and fired.

Not at the naked stranger stumbling to his feet.

At Janice.

Her body was flung against the headboard. There was a terrible silence, followed by a voice saying, "No! No! Please!" This was the stranger, cowering on the floor as the pungent smell of cordite hung in the air.

Then, as if startled out of a trance, Sheriff Harrison did something that made no sense from one perspective, and perfect sense from another. He rushed over to his wife, gently wrapped her in the bloody sheets, carried her out to his car, and drove, siren blaring, to the nearest emergency room.

I first heard this story of the sheriff and the shooting from a colleague, and though I've changed their names, I've included other details of the story relayed to me to illustrate a point that has been emerging in my research on infidelity in humans: Betrayal offends all the emotions, and our responses are often rooted in an unbridled instinct to defend our pair bond.[1] As many of us know from our own lived experience, jealousy can make us act in ways we would never have guessed ourselves capable of.[2] And the science of why this happens is still unfolding.[3]

Only 3 to 5 percent of mammals, and 15 percent of our fellow primate species, exhibit mating patterns — and associated social behaviors — characterized as social monogamy.[4] Still, species like the bald eagles that roam North America, the nocturnal owl monkeys native to cloud forests of South and Central America, baleen whales deep in the polar seas, and shingleback skinks in the desert grasslands of Australia will scream, bite, and claw to make sure an interloper doesn't get too close to their pair-bonded partners — even, sometimes, at the expense of those very partners. Observational studies and genetic paternity analyses have shown that few species that are socially monogamous are also exclusively sexually monogamous, so most socially monogamous animals attempt to enforce their mate's sexual fidelity.[5] In the natural world, this behavior is called "mate guarding," though in humans it typically manifests in less-explosive ways than what happened with Sheriff Harrison.[6] We've all seen that guy at the party who won't let his boyfriend out of his sight and gives other men the stink eye if they get too close, the girlfriend who consistently talks trash about her boyfriend's female colleagues, or the husband who tends to insert himself between his wife and any perceived potential rivals.

These are all strategies that researchers have documented and analyzed — as in one study showing that men who exhibit mate-guarding behaviors the most tend to have poorer-quality sperm.[7] Mate guarding, especially if it is excessive, can indicate insecurity about status within the relationship. And when mate guarding turns into aggression, jealousy, or paranoia, our relationship can quickly become toxic.

At the same time, mate guarding is also adaptive — it is a signal of investment in the continued success of the pair bond.[8] In fact, some female African elephants will actively *encourage* mate guarding from a male she's mating with during the relatively short period she's in estrus (i.e., sexually receptive), to prevent all the other bull elephant males from incessantly chasing after her.[9] Something similar often happens with humans: A little jealousy from our partner can be attractive to us, because it indicates that they take our relationship seriously and don't want to lose it.[10]

Humans are wired to be jealous, to protect pair bonds once they have them, which is likely why we have strong opinions about infidelity. A Gallup poll from 2025 found that 89 percent of Americans consider infidelity immoral (compare that to divorce, which 19 percent consider immoral); this number has remained relatively stable over the quarter century that Gallup has surveyed Americans about perceptions of morally wrong behaviors, and it is consistent with their trended data from three years prior, when 89 percent of respondents also believed that extramarital affairs are immoral (though a lower rate of 13 percent considered divorce immoral that year).[11] Given that our biological drive to create long-term bonds is so often deeply at odds with our sexual impulses, the questions become: Why do some people set their sights beyond the pair bond? How do people define infidelity? Are some of us simply hardwired to cheat, and is there an evolutionary advantage to doing so?

GENETICALLY UNFAITHFUL

According to an apocryphal story, the thirtieth president of the United States, Calvin Coolidge, and his wife, Gracie, were on a tour of a government farm when at some point they separated, and the First Lady was taken to an area of the farm where she noticed that one rooster was getting a lot of chicken attention, as it were. When she asked if that was normal, she was told that the rooster would often mate dozens of times a day. "Tell that to the President when he comes by," she is said to have replied. Upon receiving her message, Coolidge asked, "The same hen every time?" and was told, "Oh no, sir, different hens each time," to which Coolidge supposedly responded, "Tell *that* to the First Lady."

While this exchange may or may not have actually taken place, the urban legend became the basis for what has come to be known as the Coolidge effect: an effect in biology where, over short intervals of time, the introduction of a new mating partner can cause a male to be willing and able to engage in sexual activity again.[12]

If you put a male rat in a cage with a female rat, after the two copulate, the male typically has a post-ejaculatory refractory period in which he will be unable to have an erection and copulate again with the same female until some time has passed. But that refractory period is shortened if a new female is introduced.[13] Human males, too, tend to have a recovery period after ejaculation before they can experience an erection again.[14] While research on whether human males experience a Coolidge effect of reduced refractory period in response to a novel sexual experience has had mixed results, anecdotal evidence suggests that they might; it's why some couples dealing with erectile dysfunction experiment with introducing novel experiences like pornography or different sexual toys.[15]

The Coolidge effect is thought to be caused by habituation of a particular sexual stimulus. That is, after repeated exposure to the same stimulus, people and other mammals tend to react less strongly — we get used to a good thing and it stops giving us the same feelings of pleasure.[16] Most people intuitively know we can habituate to food, becoming bored with the same meal over several days. It turns out that a very similar process happens with sexual arousal. Just like with food, this doesn't mean we are unable to eat and enjoy the same food items, but we'll enjoy them even more if we switch around the presentation, preparation, and pairings with other food items. In other words, sexual habituation doesn't necessitate finding a new intimate partner, but it does encourage finding new ways to introduce novelty and variety with our existing partner.

For years, research — including one massive study led by David Schmitt of 16,288 people across ten major world regions — has shown that, on average, men prefer sexual variety more than women, and subsequently are more likely to commit infidelity (generally defined by researchers as sex with someone other than one's exclusive primary partner).[17] The old cliché — that men are wired to spread their seed as far and wide as possible, while women evolved to be more choosy, so as to find the most genetically fit partner — is an oversimplification at best. Yes, on average men have more interest in sexual variety and are more likely to stray than women, but newer studies suggest that this gender difference might be attenuated by a variety of social and cultural factors.[18]

In fact, many of the differences in behavior that seem gendered often have more to do with the norms and structures of our cultural and political environments than anything necessarily innate to our biology.[19] For example, in patriarchal societies where uncommitted sex has long been stigmatized for women, women's behavior may be shaped by feelings of shame or fear of social disapproval; these concerns largely don't exist for men, for whom having uncommitted sex is viewed as a sign of virility.[20] More recently, after the June 2022 U.S. Supreme Court *Dobbs* decision overturning *Roe v. Wade,* our research has shown that women are more cautious about their sexual behavior, because they choose not to risk an unplanned pregnancy that they may not be able to terminate.[21]

We are constantly adapting our behavior in response to ever-shifting cultural contexts. In societies where women have less legal authority and gender equity, for instance, their responses to betrayal tend to be more calculated; they are more likely to try to work to maintain the relationship rather than ending it or overtly punishing their partner.[22]

Even though the question I'm most frequently asked is why people engage in infidelity, a close second is how many people cheat or stray from their primary relationship. Studies on infidelity are all over the map in terms of methodology, making this a tricky question to answer. Some of this variability depends on what time period is being studied — are we talking about infidelity over the course of a lifetime, the last year, or in your current relationship? And then there's the question of what specifics are factored into the definition of cheating. Are we talking about kissing someone who isn't your partner, having sexual intercourse, or sharing the kinds of personal disclosures that tend to characterize an emotional affair? What is the cultural context for which we are asking and measuring? These unspoken variables make the question of how many people cheat a difficult one to answer. The North American statistics we have suggest infidelity occurs in anywhere from 5 to 90 percent of the population, depending on the study.[23] That is an uninterpretable range. If we're defining infidelity as a sexual betrayal, we could reasonably narrow that down to 20 to 50 percent, which is still a pretty huge range.

In 2010, my research program unearthed an astonishing discovery: that some people have a genetic predisposition to engage in infidelity, an inherited trait that became known as "the infidelity gene." This research on the genetic bases of infidelity is the kind of study in which we take a gene whose function is already known (in this case, DRD4 and DRD2, genes that influence the function of the brain's dopamine system) and investigate how it maps onto variation in human behavior (in this case, sexual behaviors and risk-taking).[24] It remains one of the most exciting studies I've done because the findings are so surprising, are so theoretically cohesive, and shed so much light on the great mystery of human infidelity.

We brought 181 young people into the lab and ran them through a battery of behavioral surveys on their romantic and sexual histories, their smoking habits and nicotine dependence, and their risk-taking and novelty-seeking proclivities.[25] First, we administered a delayed reward discounting test to determine how impulsive or measured each participant tended to be. This task, the monetary choice questionnaire, is a thought experiment that goes something like this: If I offered you $10 today or $20 next week, which would you choose?[26] What if I offered you a choice between $20 today and $100 in six months? Researchers can adjust the amounts and the time differential to home in on someone's tendency toward impulsivity.

Once we had results from that task, we could control for impulsivity when we dove into an analysis of participants' behaviors and genetics. We then asked participants to swirl around a small amount of mouthwash containing alcohol and spit it back into a sterile tube to collect a sample of a few inner cheek cells to be used for DNA extraction. Once back in the laboratory, we processed the samples and took a careful look at each individual's dopamine receptor genes.

When we think about our everyday experience of dopamine, we often think of how it feels good when we do something that floods our brain with a sudden surge of this potent neurotransmitter: Jump out of a plane? Spike of dopamine. Use certain drugs? Spike of dopamine. Genitals get stimulated? Spike of dopamine. The rush of dopamine is due to anticipated rewards, the excitement and uncertainty of what we expect to happen. Dopamine is, in this way, also involved in risk-taking and sensation-seeking; one study in people with Parkinson's who were treated with a pharmaceutical drug that increases the amount of dopamine available to the brain found that around one in ten actually developed a gambling problem as an unintended side effect.[27]

To understand this clearly, let's cover some basic genetic principles. Genes are responsible for variations in traits expressed in an individual's physiology (for example, blue eyes versus brown or green eyes). Genes come in pairs, and we generally inherit one from our mother and one from our father. Modern genetic research has shown that most of our traits, from our physical morphology (e.g., height, weight, breast size) to our behavior, are what we call polygenic, meaning that many genes are collectively contributing to the trait and its expression. But some genes can have more influence than others. Specific genes also vary, with different "versions" at particular points on the gene, and sometimes these individual differences can be measured. In the case of the dopamine genes we were studying, we could assess whether individuals carry a "long" or a "short" version of a part of the gene. While a gene is a central part of DNA structure, alleles are the different variations of that same gene, coding for diverse expression of a trait.[28]

In our study, we were interested in the genes that determine how our brain's receptors take in dopamine.[29] Past research has linked people who have the dopamine receptor long alleles to sensation-seeking behavior, and therefore more novelty-seeking and risk-taking.[30] One explanation is that people with long alleles on their genes have a neurobiological setup that is a little like a clogged garden hose; in order to get the same dopamine high everyone else gets, they have to flood a lot more neurotransmitter through their hoses.

When we then compared people with short alleles on those dopamine receptor genes to those with long alleles on the same gene, we found that people who have the longer version of the dopamine D4 receptor gene (DRD4) were 50 percent more likely to report having engaged in infidelity. Furthermore, among those in our study who reported engaging in infidelity, the individuals with the longer alleles were much more likely to have strayed multiple times. In other words, those with the sensation-seeking variant of the gene were more likely to have committed infidelity in the first place, and when they did, they did it more often.

While these findings pointed to a clear genetic component of cheating, I cannot overstate how much this finding does *not* mean that if a person has a certain genetic makeup, he or she is somehow destined to cheat. Rather, what we found is evidence that some people might be biologically impelled to seek out novelty and thus take more risks, which may or may not include the novelty and risks associated with infidelity. Our genes influence the tendencies, predilections, and desires we all carry with us into our intimate lives, but biology is not destiny. While our study revealed statistically significant differences between the two groups, we also found that plenty of people with the gene didn't engage in infidelity, and plenty of people without the gene did. Our brains evolved and grew in size over the past few million years precisely to provide us with remarkable decision-making capacity, which makes us free agents able to make informed choices; while those choices may be heavily influenced by our biology, they are not uniformly dictated by it.

Unfortunately, despite our paper's exhortation cautioning against overcharacterizing our findings as "the infidelity gene," that is exactly what it came to be called in the popular press. In hindsight, if I could do this all over again, I perhaps would have fought the media's characterization a bit harder. Also, where the academic paper used the behavioral biology term *promiscuity* to describe sex outside of pair-bonded relationships, today I would probably use the phrase "sex outside of the primary pair-bond relationship" or "extradyadic sex." Not as catchy for clickbait headlines, but when it comes to scientific research, more-precise language is always better for our ability to interpret, replicate, and understand how new data may impact our lives.

You might be wondering what possible adaptive advantage genes associated with infidelity could confer. One possibility is that a population would benefit from having some — but not all — members with greater sensation-seeking traits, in a sort of distribution of behavioral temperaments. Put more simply, we want a balance of risk-takers and non-risk-takers in our population. Shedding any one gene entirely — like the so-called infidelity gene — might also throw a significant amount of baby out with the bathwater.[31]

So while we don't want everyone in a population to be predisposed toward sensation-seeking and risk-taking behaviors, we don't want to eliminate this variation entirely, because there is something about these kinds of behaviors that we need as a species. This is known in evolutionary biology as frequency-dependent selection, meaning that evolution maintains a genetic variation because of the relative benefits of trait diversity in a population.[32]

Consider, for example, the bluebird. There is a subset of the species that is sensation-seeking and highly prone to novelty, and another that is more neophobic, meaning they prefer what is familiar. At the population level, especially for social animals, we want some individuals who are going to stick to the nest and protect the home front, *and* some who will bravely venture to the other side of the mountain and look for new resources, new foods, new territories, or new mates. But we don't want too many explorers, because exploring is risky; there very well could be many new high-quality foraging opportunities in the great beyond, but there could just as easily be a wasteland or a colony of predators on the other side of that mountain, and if you never come home, you can't pass on your sensation-seeking genes.[33]

This is analogous to human history in terms of our ancestors' exploration across the globe; many ships never made it to a destination or never returned, but those that did facilitated the exchange of spices, medicines, resources, and ideas. Humanity's migration across the globe was pivotal to our explosion as a species, and we became arguably the most dominant, most widely dispersed animal on planet Earth precisely because we have these behavioral variations that spurred exploration by some humans. Early work on DRD4, the gene implicated in rates of infidelity, shows that the long allele, the sensation-seeking variant, likely evolved about 50,000 to 100,000 years ago, and if we think about what *Homo sapiens* was up to at that moment in time, it aligns with the period when our species geographically dispersed. Anatomically modern humans began migrating from East Africa roughly 50,000–70,000 years ago, making it to Asia some 50,000 years ago and to Europe 40,000 years ago.[34]

Then and now, our ancestors who were the most likely to explore new territory might, by the same genetic stroke, also be more likely to cheat.

WHO'S YOUR DADDY?

In some colonies of western gulls in the Channel Islands, where the females outnumber the males, the gulls tend to form female-female pair bonds in order to cooperatively rear their young.[35] In same-sex pair bonds, parental certainty ceases to be a tool of relationship negotiation. In different-sex bonds, however, there is an alternate calculus, as the female of a species is always the one with parental certainty, which is obviously a valuable asset.[36] Fathers have a less certain role; cross-species studies on genetic paternity in socially monogamous animals tell us that a mother is not always pair-bonded to the genetic father of her offspring.

Among heterosexually bonded humans, genetic paternity studies have revealed some Hollywood-style drama. In a now classic study exploring worldwide nonpaternity rates, it was found that 9 percent of babies were birthed to women who were not partnered to the natal father (meaning the new fathers who showed up to the hospital were not the actual genetic father in 9 percent of cases observed). This study sometimes gets sensationalized with declarations like "One in ten kids are raised by men who don't know they aren't the biological father!," which is not particularly useful to understanding what's really going on.[37] It turns out that many dads don't actually think they are the birth father, and if we control for "paternity confidence" (as biological anthropologist Kermyt Anderson did in one excellent cross-cultural analysis), the numbers are something closer to 1.5 to 2 percent of dads who are actually being duped.[38]

In evolutionary terms, this is still a high enough misattributed paternity rate to have a significant impact on the genetics of the next generation, especially if the genetic fathers are siring offspring with multiple partners, distributing their DNA across multiple nests.[39] In other words, in a large population, 2 percent can magnify to thousands or millions of pregnancies. What this means for our species in terms of infidelity is that in spite of powerful adaptations for social monogamy, a genetic predisposition for novelty-seeking and risk-taking (even taking into account that the genes being passed down are not always expressed) is occurring in enough of the population for the adaptation to sustain itself and be passed down to the next generation.

But much like other forms of risk-taking, infidelity can be extremely perilous for those who are caught — especially for women, who have historically borne the brunt of the consequences, losing everything from their current relationship and financial stability to custody of their children and even, in extreme cases, their lives.[40] So why transgress the pair bond? We've seen the evolutionary advantage to infidelity at the population level, but for it to continue despite the risks, we should expect to see the behavior result in some genetic benefit for individuals as well.

In some cases, the benefit may be to facilitate finding a new relationship partner and therefore satisfy our species's craving for variety. In others, the evolutionary advantage may be more offspring, or more diverse offspring. It may be advantageous to have more varied descendants with different genetic mixes, even possibly being raised in different nests.[41] Think of it this way: If the nest your backyard bird has with its partner gets hit by a predator, they might lose all their young that breeding season, but if one of them happened to also sire offspring in their neighbor's nest, their genes might still make it into the next generation. Or, in another real scenario from the natural world: If your offspring are genetically susceptible to a new pathogen that only affects those with particular immune function genes, the offspring you have with a different partner have different genes and might have a greater chance of surviving a plague. In these cases, there is enough of an evolutionary advantage that sex outside the pair bond is adaptive. But the advantages also vary by sex and gender, and are critically dependent on the ability to solicit additional parental support. For women, sexual activity leading to reproduction with someone to whom she's not pair-bonded may be most advantageous when she's pair-bonded to someone else — if her pair bond survives the infidelity, she has diversified the genetics of her offspring without losing the nurturing advantages of the pair bond. If the relationship does not survive the infidelity, however, the female gains fewer advantages by reproducing outside the pair bond, because she now must take on all of the work of nurturing offspring herself.

This subtle but significant detail is what got me interested in sexual hookup behavior. From an evolutionary perspective, we don't generally expect a pair-bonded species to have uncommitted sex unless they are already committed to another, in which case that uncommitted sex is also infidelity. In contemporary sexual hookup culture, it tends to be single people having uncommitted sex, as opposed to partnered people having uncommitted infidelity sex.[42] In the case of partnered people, this raises an altogether different set of questions: When people do engage in infidelity, to what degree is their sex beyond the pair bond uncommitted versus a longer-term affair; moreover, what are the potential evolutionary consequences of this distinction?

Infidelity is an act that puts the sex drive and the intimacy drive into direct and obvious competition. That's why it can be so thrilling, so confusing, and often so destructive — particularly when infidelity involves long-term, enduring relationships beyond the pair bond rather than simple one-off sexual encounters. It is my belief that the intimate bonds forged by enduring affairs have had a long-overlooked impact on our species's evolutionary history. This is the shadow life of intimacy, in which we see an even greater collision between the two drives.

BUT WHY WOULD *I* CHEAT?

As with most of what goes on in our intimate lives, the reasons people commit infidelity are incredibly complex. In one study with my collaborator Dylan Selterman, we looked at the primary infidelity motives in a largely heterosexual sample of 495 young adults (we left the term *infidelity* undefined and allowed respondents to decide what it meant to them), and ultimately identified eight distinct reasons people cheat:[43]

1) Anger ("My primary partner had been unfaithful to me, and I wanted to even the score"; "Before my affair, my primary partner and I got into an argument, which led me to seek revenge")
2) Sexual desire ("My primary partner had lost interest in sex"; "My primary partner refused to perform certain acts during sex that I normally enjoy")
3) Lack of love ("I was not sure if I really loved my primary partner"; "I was not sure if my primary partner was the right person for me")
4) Low commitment ("I was not very committed to my primary partner to begin with"; "Even though we were seeing each other, we were not technically 'in a relationship' publicly")
5) Esteem ("I wanted to boost my self-esteem / feel better about myself"; "I wanted to assert my independence and autonomy")
6) Situation ("I was drunk/intoxicated and was not thinking clearly"; "I was overwhelmed at the time due to external stressors — e.g., school, work, family issues")
7) Neglect ("I felt neglected by my primary partner"; "My primary partner was emotionally distant")
8) Variety ("I wanted a greater variety of sexual partners"; "I am the kind of person who cheats; it is a part of my personality")

In a separate follow-up study, we examined how these motives impacted infidelity experiences and outcomes. Perhaps not surprisingly, we found that people who strayed for reasons like lack of love and neglect enjoyed more intimacy in their affairs: going on dates, engaging in public displays of affection, and being told "I love you." These affairs also tended to last longer, whereas affairs arising out of situational motives were more likely to be brief sexual encounters. In other words, people who cheated because they weren't getting the love they needed from their primary relationship weren't cheating for sex; they were cheating for *intimacy*. Although sex was a factor, in some cases it may have been merely a vehicle for an intimacy goal or even a byproduct. Interestingly, the closer people felt to their primary partner before the affair, the less emotional satisfaction they derived from the affair, suggesting the closeness of a primary relationship may get in the way of forming deep emotional bonds with another partner.[44]

In terms of gender differences, we found that men were — perhaps unsurprisingly — more likely to report infidelity motivated by sexual desire, sexual variety, and situational factors, whereas women were more likely to report infidelity motivated by lack of love or neglect. In other words, men were more likely to cheat because they had an opportunity or wanted to experience something new and different, and women were more likely to cheat because they felt unloved or unwanted by their primary partner. Men seemed more motivated by the desire for sex, while women were more often looking for intimacy.

When it comes to being cheated *on,* research shows that women are generally more reactive to their partner engaging in emotional infidelity, while men are generally more reactive to their partner engaging in sexual infidelity — although when given a third option, most people, regardless of gender, are reactive to both.[45] In real life, emotional and sexual infidelity can also be a false dichotomy; often, it can be hard to separate one kind of betrayal from another. Both hurt, sometimes a lot.

As a landmark series of studies by leading evolutionary psychologist David Buss showed, relationship history mattered, too. In Buss's studies, men who had been in committed romantic relationships in the past reported feeling more distressed by the *idea* of sexual infidelity than those who hadn't. Men who had experience with entering the social contract of a relationship were more upset when that contract was broken; they knew what it was like to have "relationship rules," so they were more upset by the idea that their partner wasn't following them. When the researchers invoked jealousy from imagined infidelity, they also found physiological effects such as increased heart rate and increased electrodermal response.[46]

Our reactions to betrayal — jealousy, anguish, sadness, anger — are both psychological and physiological.[47] There is so much at stake with either kind of infidelity — evolutionarily speaking, women stand to lose access to resources and/or paternal investment, whereas men lose their sense of paternal certainty. And for both men and women there is the potential to lose their partners and all the ups (and downs) that come with pair bonds. Taken together, it's easy to see why humans respond to infidelity so viscerally and often violently.

WHEN CHEATING ISN'T CHEATING

For years, my friend Rachel enjoyed what she described as an "open marriage." Early on in their relationship, she and her husband would sometimes pick up a third to come home with them for the evening — mostly because it excited him to watch her have erotic pleasure with someone else. As they settled into their marriage and began raising children, his interest in threesomes waned, but her desire for other men remained. Since he spent so much of the year traveling for work, they settled into an unspoken agreement: So long as she kept her dalliances discreet, they would operate under a don't-ask-don't-tell arrangement.

Then COVID hit.

Suddenly her husband was no longer able to travel. He was at home for months on end, and it became increasingly difficult for her to adhere to the parameters that had always kept feelings of betrayal and mistrust from seeping into their relationship. Now she would sneak a meeting with one of her flings while out running errands, or when her husband was home with the kids, and found herself lying to him about where she was going and why, making up excuses to get out of the house, and engaging in all kinds of machinations to cover her tracks. Suffice it to say, their nonmonogamy had lost the consensual element, and the relationship was starting to suffer.

Definitions of infidelity within a pair bond are important, something I too learned the hard way in my own personal life. While in graduate school — studying the evolution of monogamy — I remember the sinking feeling of finding out that a woman I'd been dating was also involved with someone else. I had unexpectedly noticed her distinctive car parked on a different side of town from where she'd said she would be that night. During the few moments while I walked to my own car in the same parking lot, scratching my head about whether to call her and see what she was up to, she emerged from an apartment building attached at the lips to a man who was definitely not the sister she'd said she was getting a bite with that evening. I nursed my wounds and complained about her behavior to a close friend — who asked whether she and I had promised to be exclusive with each other... and...we hadn't. That moment called some of my assumptions about fidelity into question.

Having multiple partners isn't, in and of itself, a problem for our intimate lives. *Betrayal,* as in the lies a partner tells to cover up an affair, is the real poison to any intimate relationship.[48] As we have seen, people in thriving consensually nonmonogamous arrangements seem to have found ways to avoid feelings of betrayal or transgression, balancing their interest in sexual or romantic variety with allegiance to the trust that is a cornerstone of a primary relationship. Some of the very newest data say that some people, such as couples with wildly mismatched libidos or partnerships that over time have turned solely companionate rather than passionate, may be better off in terms of mental and sexual health by choosing nonmonogamy.

For a negotiated consensually nonmonogamous partnership to succeed, there typically needs to be a set of rules around their fidelity and what would constitute a transgression.[49] It might be off the table to develop romantic feelings. Sleeping over might be a no-no. Or dates with other partners might be allowable only on certain nights of the week. And preferences around information sharing should always be respected, as in a situation where one partner does not want to know the details about sex with other partners, or even to know if something's happened at all. Whatever the specifics, having agreed-upon parameters helps limit feelings of jealousy or pain. When both partners know the rules, following them becomes an act of mutual respect — a way to deepen the relationship, instead of weakening it or casting it aside.

HIGH FIDELITY

Given the biological, psychological, and cultural reasons to commit infidelity, my colleagues and I have recently begun to pose the opposite question: What are the reasons people *don't* cheat?

Our research is ongoing, but our early thinking suggests that it may very well come down to empathy — the same neurological function that allows for deep connections with others, part of the very foundation of our capacity for intimacy. Most of us know what cheating could do to our partners, to our other relationships, to our reputations. Cross-culturally, even in places with high rates of infidelity, betrayal is still widely understood as a relationship transgression that partners generally attempt not to commit.[50] And most people in long-term committed relationships don't ever engage in infidelity. It's hurtful to be the one "cheated on," and in loving relationships we generally try to prevent hurting our partners, whether directly, as in the case of a betrayal, or indirectly, perhaps by causing them shame or doing something that upends a shared friend group or other interpersonal relationships. It's interesting to note that infidelity is linked to what's known in psychology as the "dark triad" of personality traits: psychopathy, Machiavellianism, and narcissism, all of which contain elements of disregard for the feelings of others.[51]

One of the cornerstones of a romantic relationship is that certain things are meant to be intimate and private — special only to you and your beloved — and this is just as true in consensually open relationships as in socially and sexually monogamous ones. For some, the thing that only the two of you share might not be sexual activity. Instead, maybe it's the bizarre nicknames you have for each other. Or the fact that someone besides you knows your exact Starbucks order, down to the soy milk and extra pump of vanilla. Or it could even be the memories that you've made and shared together.

I once had a girlfriend who would cuddle close to me, sweetly hold my face, and tell me, "I love those bright blue eyes." In those intimate moments cuddling together beneath the blankets on cold upstate New York mornings, feeling like I was her most special person on the planet, I understood what it feels like to be delighted by another person — and to want to protect them and the relationship.

But I also understand that the same intimate bond that ties one partner to another can often lead them to cause the other person pain. The problem is that when we hurt the ones we love — even when, like Sheriff Harrison, we do it *because* of the love we feel — we often rupture the very bond we mean to protect.

Extra-pair copulation (as infidelity is so sexily termed in the animal behavior literature) is present in almost all pair-bonding species biologists have studied, and it often leads to violence. I vividly remember a time I went on a birdwatching hike through the Binghamton University nature preserve in Broome County, New York. What I witnessed was a particularly violent cuckolding episode. I had my binoculars trained on a pair-bonded couple of red-winged blackbirds when suddenly another male came swooping in through the reeds toward the pair.[52] It happened so quickly I never did figure out which one was the interloper. The two males had their talons up, grappled, fell through the reeds, and tumbled into the swampy water. Only one male flew away. Aggression and impulsivity are bound up in sexual betrayal for any animal, no question. And with humans, the unfortunate reality is that this violence tends to be inflicted disproportionately on women.

At the same time, humans, unlike most animals, have many other, nonviolent means of repairing a relationship in the wake of a betrayal. While we're not the only species to stray from the pair bond, we are the only one with the capacity to reframe what infidelity means to our intimate relationships, the only one with the ability to see a betrayal as a signal pointing us to issues in the relationship or within ourselves that we can explore.[53] Maybe it means a breakup, maybe it means testing an open relationship, or maybe it means working on some area where intimacy was lacking.

The science is clear on one point of reflection, however: Just because someone strays, that doesn't mean they don't love their partner. It may mean, though, they've eroded the trust that binds that relationship together. Ultimately, trust is at the heart of what it means to be bonded to another person.

I've spoken to so many people who insist they would never in a million years stay with a partner who cheated; yet, in study after study we see that more than half of couples do stay together and try to work through the resulting turmoil. What doesn't explode a relationship might make it stronger, and the intensity of those emotional bonds goes a long way toward explaining why we stay when our partners stray.

But what happens to our brains and bodies when the pressures on a pair bond cause it to break? And what is an intimate animal to do when love is truly lost?

Chapter Eight

BREAK

When my friend Jessica broke up with her boyfriend of five years, her closest friends mobilized and showed up at her house in waves. There wasn't much we could do to "fix" her broken heart, but we could listen, empathize as she processed her emotions, and let her know she wasn't alone. When I arrived for my shift on day three of the breakup, she was curled up in a recliner, wearing a baggy sweatshirt and leggings she had clearly slept in, her eyes swollen and her face streaked with tears.

Though she and her ex had met when they were both in graduate school on the East Coast, he now lived in California and she in Indiana, both of them beholden to the trajectories of their careers as academics. After over a year and a half of physical separation, the two had become masters in staying connected — streaming movies together, sharing playlists through music-sharing apps, snatching a few minutes on the phone in the hallway of this or that hotel conference center, their text thread a transcript of love notes, peppered with strings of emojis and funny GIFs.

But...

"If we lived in the same city," she told me through a scrim of tears, "if we could fall asleep together and wake up together... there wouldn't..." She struggled to find the words.

It was at this point I realized the mug of tea I thought Jessica was drinking was, in fact, vodka.

Working at the Kinsey Institute exposes me to the ways in which people of all demographics and from all over the world conduct their romantic and sexual lives, including why and how couples break up. Even happy couples argue, and when they do, they argue the most about intimacy (showing physical affection, frequency of sex) and leisure (going on more dates together, having more fun together), but they argue for the longest amount of time about the household (being organized, doing chores, consulting each other on household decisions). On average, singles have had three difficult breakups in their lives (except for the poor singles who happen to live in New Hampshire, who in our study had an average of eight difficult breakups in their lives — causation unclear, though one has to wonder whether it might have something to do with their state motto, "Live Free or Die").[1]

One phenomenon we are starting to observe is the growing number of partners who still live together, often for economic reasons, after the pair bond dissolves. In one poll of those living in London, where the cost of living can be quite high, one out of three respondents had continued to live with an ex after a breakup, and of those who did, seven out of ten continued that arrangement for more than three months.[2]

One couple I met not only continued to live together after they separated but even continued to share a bedroom and queen bed, in which plenty of snuggling continued to happen, but they stopped having sex with each other. Essentially, they had reconfigured their relationship to remove sex and give permission to each partner to date other people, though neither could actually bear to date anyone else. I remember asking them, separately, about what they still did and didn't do together: sex (no); dinner (sometimes); cuddling (just in bed at night); showering together (no...well, maybe a few times). They eventually got back together. Were they ever really broken up? Sort of. One problem with interpreting this arrangement is that there's no clear way to define these types of messy realities that people experience in relationships. They weren't together, but they weren't apart, either. Often, we entertain a fantasy that a clean break is the solution, perhaps the only solution, to ending a relationship. But reality is sometimes, if not usually, a bit more nuanced than that. The human animal finds ways to get its intimate needs met regardless of comfortable categorization.

Loss of love is not necessarily limited to the confines of a romantic breakup, of course; people die or move away, or bonds are otherwise broken.[3] We've talked about infidelity as a risk of entering a romantic relationship, but the reality is that love *itself* is an enormous risk. Breakups are just one quantifiable way we can try to examine the risks of intimacy in the lab.

Post-relationship grief is a form of bereavement, one that is as much about the end of the life you once shared together as it is about the loss of the partner.[4] That loss can be painful even in cases where we might be happy about a relationship ending and see it as a positive turn of the page.[5] In several studies of relationship dissolution, researchers found that for many the emotional pain can be characterized as meeting clinical criteria for mild to severe depression, and in some cases individuals suffer from symptoms consistent with post-traumatic stress, ranging from psychological intrusion to avoidance.[6] In one epidemiological study of adolescents, romantic breakups were a major predictor of first onset of a major depressive disorder, and distress from a relationship breakup is a major risk factor for suicide, especially among young people.[7] In reflecting on this data, I often think of a colleague of mine who described the phone call she received from her teenage son after his high school girlfriend broke up with him. The pain in his voice was so palpable, her first thought was that he had been in some kind of accident.

Grief can be exacerbated by loneliness — and vice versa — especially among older adults. And, as we've explored, loneliness research has shown that the physiological effects of being alone in turbulent times can include increases in reported levels of suffering and may make periods of mourning last longer. This is why it's important to have friends, social networks of support, even pets that help pick us up. There is a science to successfully navigating post-relationship grief.

I have long wondered why in Western developed societies we don't give ourselves more room to mourn the loss of a relationship. Is there a better way to grieve, even as we try to pick up the pieces and move forward with our lives? We might start by having more compassion for the recently broken-up, as we do for those who lose a partner or family member because of death. The Jewish tradition of a week-long period of sitting shiva with family and friends after the death of a loved one might be a good model for mourning a relationship loss, but in the absence of a formal ritual we might take inspiration from an example that showed up in my social media feed recently when a friend posted a "memory" photo from five years earlier that the platform had suggested from her prior posts. The memory was a postcard-like image of my friend with her late wife, who had passed away from cancer, the two of them staring lovingly into each other's eyes, backdropped by a beautiful sunset. Down in the comments, I witnessed something beautiful. My friend's *current* wife had posted: "I love this photo. I never got to meet her, but I know how special she was <3 <3 <3."

It's easy to dismiss a breakup as something frivolous, but the data show there's a very real cost in terms of human wellness. We lose weight. We gain weight. We get swept up in a roller coaster of mood swings: sad, angry, relieved, sad again.[8] In one study, my colleague Helen Fisher found that some people even feel physical pain during relationship loss.[9] And the toll of all of this is exacerbated in those who are struggling with a serious health issue or medical condition. What happens if, for example, you're taking medication to manage a serious condition, and your breakup causes such significant stress in your life that your medical adherence becomes uneven or nonexistent?[10] And what happens if you're someone who is struggling financially as well, and suddenly you need new housing, or a different arrangement for living, working, or access to healthcare?

Many of us have been subject to the ridiculous exhortation to "get over it," which is not a thing that can be done on command. It's just bad advice. What we need instead is compassion. Only by acknowledging and honoring the intensity of the grief that comes with the decoupling of a pair bond can we truly help our friends and loved ones, and ourselves, to heal, feel whole, and regain their capacity to love again.[11]

WISH I COULD QUIT YOU

Trying to move on from a relationship is, scientifically speaking, very similar to trying to kick a drug habit. Neurobiologically, passionate love is much like craving a substance that gives us a euphoric buzz, and it turns out to be one we can get addicted to.[12] Fisher has gone so far as to argue that, in fact, love *is* a form of addiction, and one that we have trouble quitting even after a relationship ends.[13] In a preliminary study, when she and her team of collaborators brought fifteen people who had been recently romantically rejected into a medical laboratory for an fMRI brain scan, they found activation of brain areas involved in cocaine addiction; in other words, the brain of someone who had been through a recent heartbreak looks remarkably like that of someone going through cocaine withdrawal.[14]

Moreover, they found that several regions of the brain associated with passionate love were also involved for those who reported still being "in love" despite having been rejected, suggesting that simply removing the stimulus (i.e., the object of one's affections) doesn't make feelings of love suddenly go away. In fact, loss of the beloved can intensify certain feelings and emotional responses, and brain regions associated with grief and pain are stimulated by such a loss.

The evolved capacity for love and sex is extraordinarily powerful and profound, a complex patchwork of motivational systems and emotions that leads us to do wonderful, occasionally ridiculous, and even dangerous or awful things. Different people experience relationship loss quite variably, and those responses also vary from one person to another. Maybe you have a friend who goes through one breakup and comes out just fine on the other side, then goes through another and sinks into depression. In the case of breakups, there isn't just one single style of grieving; just as degrees of intensity vary with each unique relationship, our response to a breakup will vary depending on the combined variety of unique relationship factors.

One of the true delights of being at the Kinsey Institute is that I'm surrounded by an abundance of brilliant colleagues who are at the top of their game. My longtime collaborator and friend Amanda Gesselman, for instance, is a social-developmental psychologist who works in the office next to mine, so when an intriguing question occurs to one of us, we can hop next door or down the hall to the conference room and start working out our ideas on a glass dry-erase board. It was exactly in this manner that we hatched a plan to study some of the dynamics around breakups.

Some of our early findings have highlighted a somewhat unexpected pattern: couples taking time to reevaluate their relationship before pulling the trigger on breaking up. In fact, our data indicate that over a third of young adults have gone through a relationship "break," during which they consciously slowed things down but did not actually end the relationship. Somewhat surprisingly, factors like relationship length, gender, and attachment style did not predict whether these people dated, developed feelings for, or engaged in sexual activity with an outside partner during the break before returning to their partnership.

We have a tendency to think that the grass is always greener on the other side of the fence. But it often isn't. Rather, as we say in relationship research, the grass is greener where you water it. For some, "taking a break" can reaffirm the strength of the pair bond.[15] With the benefit of a little perspective — and perhaps a reminder of how hard it is out there on the open market — many will return to the relationship more motivated than ever to invest time and intimacy in it.

We already knew that many people can exhibit extreme behaviors when a relationship ends, sometimes harming themselves, their previous partner, or even others.[16] This pattern of relationship breaks suggests that relationship dissolution and heartbreak take a vastly greater toll on our mental, physical, and emotional well-being than researchers, clinical healthcare providers, and our societies have appreciated or cared to address. This is likely especially the case in our hyperconnected digital lives today, where we often feel haunted by our exes.

One of my greatest research goals is to use the science of human intimacy to normalize the idea of heartbreak as a serious threat to our physical and mental health.

MUST BE LOVE ON THE BRAIN

Behavioral scientists and relationship experts have amassed a considerable body of research related to the "on-ramp" of courtship processes (the *search*, *date*, and *mate* domains of the relationship life cycle). But by comparison, much less is known about the bumpy "off-ramp," when people *exit* romantic and sexual relationships, willingly or otherwise. As my colleagues and I have begun to dive deeper into this issue and collect more data, one thing we've tried to do is to apply more sophisticated statistical analyses in hopes of synthesizing that data into a theory that might shed some light on the complexities involved in relationship dissolution. To that end, one of our lab's studies examined the ways in which people represent the ramping down of their relationships, specifically on social media.

When we asked single adults in the United States how they manage exes on social media, we heard that nearly half of single men and three-quarters of single women "unfriend" their ex after a breakup, and more than half of singles even block their exes from seeing their social media. Most singles agree that a clean break with an ex on social media goes a long way after a breakup, with over two-thirds saying they will delete all or some of the photos of their former partner from their social platforms. And only half of participants said they reach out to wish an ex happy birthday on social media.[17]

At the same time, in the digital age, our intimate past lives can live on in ways that make it harder for some of us to move on; nearly half of men and a third of women will hold on to sexually explicit photos of their ex after a breakup.[18] And sometimes staying connected can reignite our remaining passions, as with our data showing that one in five singles has gotten back in touch with a former partner after their ex liked something they posted on social media.[19]

Generally, this is a bad idea. After all, one wouldn't advise bringing an alcoholic to a bar, or offering heroin to a recovering drug addict. We would never intentionally taunt someone in recovery with stimuli related to their addiction, yet every time you check up on an ex on social media, you're basically doing just that to yourself. Addicts in recovery are advised to remove stimuli associated with their using, to get rid of the paraphernalia; we can apply an analogous logic to breakups — stop sleeping in that sweatshirt that smells like your ex or stalking their social media. If you're not ready or willing to completely unfriend your former flame, you may want to at least turn off notifications for your ex's account so you're not receiving constant updates of his or her life without you.

Behavioral theorists have developed various hypotheses to explain why we can't stop ourselves from staying connected to or checking up on our exes. More and more, I have come to believe that romantic love makes an indelible imprint on the mind, and when a breakup occurs, it leaves a scar. I mean this both metaphorically and literally — as we've seen in fMRI studies conducted by my colleague Bianca Acevedo and others, specific patterns in brain activation associated with long-term romantic love suggest that these relationships leave residual marks on the brain.[20]

The adaptive reason humans evolved a proclivity for such intense pair bonds is that they help us weather the unpredictable turbulence of the world, including sickness and other challenges. But the consequence of intense bonds is that it's difficult, sometimes nearly impossible, to fully shake ourselves free of them. Once we have loved someone, a little bit of them stays with us forever. In many ways, this is nature's price for intimacy.

I have a friend who can't smell fresh lavender without it triggering the bittersweet memory of a two-decade-old teenage crush, and another who gave up cigarettes years ago but still loves the lingering scent of smoke on other people's clothes because it reminds him of an ex.

The bonds we forge with a romantic partner don't get wiped away when the relationship ends, as evidenced by the stunning finding that half of the participants in one of our studies said that they have had an on-again off-again relationship at least once in their lives.[21] Our research team has come up with a number of compelling practical reasons we might return to an ex: they're a known quantity; it doesn't add to our "body count" of partners; it's easier than reentering the mating market; a part of us still loves them.

Fundamentally, I believe this form of relationship yo-yoing is a byproduct of the strength of pair bonds. It's similar to why we see so much recidivism when it comes to drug use: Our partners get us high on romantic love, which makes us feel good, and we become habituated.[22] Then, when we lose access to the source of that rush, we go through withdrawal while we relearn how to live without them. And many of us won't make a clean break on the first try.

I once interviewed a young man who hadn't quite managed to fully break away from a relationship he had cherished but which wasn't fully meeting either his or his partner's needs. Although he and his ex-girlfriend loved each other deeply, ultimately it became apparent to both of them that she was a lesbian. When they ended their romantic relationship, they vowed to remain close, and a year later they had stayed true to that promise. The friendship they forged in the aftermath of their breakup saw them through their initial sadness over the death of their romantic love, as well as the subsequent jealousy they struggled with as they started seeing new people — but it may also have hampered their ability to truly move on. Whenever they introduce a new partner into their lives, they make it clear from the beginning that their weekly Sunday brunch dates are nonnegotiable. It's made for an imperfect start to some of their new romances, and I can't help but wonder how this will play out down the road as one or both of them become enmeshed in a more dominant relationship.

One interesting finding from this body of work is that while relationship dissolution is often challenging regardless of gender, women may find it somewhat easier to break free from the emotional bonds of a past relationship than men.[23] We find this to be true at least in heterosexual marriages, where women are more likely to want and initiate relationship dissolution than their husbands; in fact, two-thirds of divorces in the United States are filed by women. In nonmarital heterosexual relationships, though, there's no gender difference in who wants relationship dissolution.[24] It may be that the role of marriage, and a woman's perceived social freedom to leave one, has changed over time. Still to explore is how these gendered patterns present themselves during relationship breakups among LGBTQ+ populations, including couples of gay men or lesbian women, as well as bisexual people in various gendered partnerships.[25] And much more research is needed to understand these patterns among those in consensually nonmonogamous relationships, as well as how relationship dissolution is experienced when someone has one or multiple other partners still available.[26] Recent studies, however, suggest that relationship dynamics including commitment and satisfaction are more predictive of relationship dissolution than gender.[27]

Evolutionarily, there is a good reason it is emotionally difficult for men and women to walk away from romantic bonds. Once we're out of that courtship stage and pulled closer and closer into each other's gravitational orbit, the attachment and commitment ties between us grow tighter, and it becomes biologically, psychologically, and sociologically difficult to uncouple — even more so when children and other family members are involved. We become entangled in each other's lives, which is perhaps a feature, not a bug, of having to pair-bond for survival and reproduction.

This is a big part of why, as we've seen, many couples decide to stay together after infidelity (and if they're married, that percentage is greater).[28] But not all relationships will ultimately survive betrayals.

People ask me all the time whether you should leave if your partner cheats. Well, if you're unhappy, if you can't trust and respect each other, it's surely worth considering. But I can't answer that question for anyone who isn't me. Experts all agree that betrayal is poison to a relationship, and that humans are great at coming up with all sorts of practical reasons to stay involved regardless: We think we can work through it, we have kids, we have a house, it was a one-time mistake.[29] And those may indeed be very good reasons for some. We say to each other, "You made a mistake this time, but I may very well make the mistake next time." Sometimes we stay quite simply because we're in love. You can love someone who betrays you; although you cognitively recognize that the betrayal is bad, your brain (and heart) is still bonded to that person. That's the piece we don't always acknowledge, and I think it's one we ought to be more compassionate about.

Relationships end for a host of reasons. One ingrained judgment that many people have is thinking about all past relationships as failures: a damaging myth I've become practically militant about dispelling. The reality is that just because a relationship ends, that doesn't mean it failed.

Sometimes the dissolution of a relationship is out of our hands altogether. A woman I know was engaged to be married in her home country, Venezuela, but her family was forced to flee the country because of political persecution. She knew she'd probably never see her fiancé again, and though young and in love, they ultimately agreed to end their romance out of practicality. Geography is a powerful force, one that is very much a part of our evolution as a clannish species dispersed across the globe.

Sometimes it's just the timing of a relationship that simply isn't right. As relationship scientist Chris Agnew and his team have shown in a series of studies, the timing of a romantic relationship is a huge factor in how it starts and is maintained: from courtship to commitment, and later to nesting—and ultimately whether the relationship grows into something deeper, or dies on the vine.[30] A person's readiness for commitment might depend on a number of practical factors, including things like where they are in their career, their education, their finances, if they plan to move soon, current or upcoming major life events, recent traumas, family factors, and more.

Other times, though, ending a relationship is a conscious decision: Something or someone hurts us so badly that staying is more painful than leaving. Most of us have watched friends and colleagues suffer in relationships where the bad starts to outweigh the good, but it might still take three, six, or twenty-four months to extricate oneself as the bad keeps piling up. These are different versions of the same evolutionary conundrum.

Psychologist Lucia O'Sullivan, who conducts some of the most innovative research at the intersections of love and sex, has shown that people often wait to have a relationship alternative before they jump ship. It's not necessarily that people are waiting for something better, or even that they plan to be with this other person; rather, humans sometimes instinctively want a safety net.[31] For humans, the cost of losing that love we feel so intensely about can make it extremely hard to pull away, even when the scales are tipped.

Sometimes you're with someone who's wrong for you but you want that person romantically or sexually all the same. You care for them, you feel passionately about them. This is another important thing I've learned from my colleagues who specialize in couples therapy. Couples might go to therapy because they're having trouble and they want to figure out how to get the spark back with their person, or keep their family healthy, or keep a business together, or maintain a lifestyle, even if they're no longer in love. In other words, there are a lot of people who are in a relationship that is in turmoil, *regardless* of their love for each other. Feelings of love and the experience of a good relationship are quite separate, albeit deeply intertwined, phenomena.

BETTER TO HAVE LOVED AND LOST

Some years ago, when I taught my course on modern love, an inquisitive sophomore raised her hand to ask the question that was likely on all their minds.

"Okay, but really: What's the *best* way to get over a breakup?"

My first impulse was to respond, "Do what I've done and listen to Adele on repeat for two weeks." Listening to sad music after a breakup may be a cliché, but there are real scientific reasons that it makes us feel better.[32] A study done by researchers in Germany found that we seek out sad music after a breakup because it delivers specific cognitive rewards: It sparks our imaginations, helps us regulate our emotions, and elicits empathy responses, all without "real-life" implications.[33] More to the point, sad songs allow us to process our raw emotions and make sense of the world when the turbulent nature of romantic love makes us feel out of control.

Similarly, that old bit of libertine advice about casual sex — "The best way to get over someone is to get under someone else" — isn't entirely misguided. After a breakup, we are mourning a loss of intimacy, and even at its most casual, sex with a relative stranger does provide *some* dose of the intimacy we need. It also provides the hits of oxytocin and dopamine that we crave — although these are cravings that we could almost as easily satisfy through travel, new experiences, time with close friends and family, massage, cuddling a pet, or exercise.

A casual fling, at any rate, is probably a better idea than beginning a new relationship before we've mourned and gotten over the old one. It can help remind us not only that there are other fish in the sea but also that some of them will want to swim with us.

Romantic relationships can sometimes feel like a game of roulette in which every spin can land on anguish of some sort. I remember calling a behavioral scientist friend one afternoon to vent about a breakup I was going through. She summed up the experience perfectly: "Sometimes it feels like Mother Nature overdid it with the emotions."

Being in love requires vulnerability, and that vulnerability by definition comes with risk. I'm reminded of the lyrics of Oscar Hammerstein from *The Sound of Music:* "A bell is no bell till you ring it; a song is no song till you sing it; and love in your heart wasn't put there to stay. Love isn't love till you give it away."

That is the risk we take in pursuit of the most fulfilling and satisfying modes of our lives. Sometimes we hit the jackpot and rake in that special potion of trust and affection combined with passion and intimacy. Every so often, Michelle and I look at each other, smile, and say "You're lucky!" and "I'm lucky!" We are amazed that we have found each other.

Almost all of us will at some point experience lost love, and for many the pain that comes with such loss. But so often we find love again. And sometimes, we land on something that feels comfortable and intimate, and that can be enough. Beyond the fires of passion, our relationships provide a bulwark against the storms and vicissitudes of life.

Chapter Nine

CARE

Jen and Dave's second child was born in November 2002. Two weeks later, on a cold Thursday night, the phone rang. It was Dave calling to say Jen needed to go to the hospital. *Now*. Could my mom come over to watch their two kids?

This was alarming but not entirely unexpected; Jen hadn't been herself. Since giving birth to Alexis (Lexi, we would call her), Jen had been forgetting things. More than once, she'd dropped a glass or something one of us handed to her — almost as if her eyesight was off. She was also having headaches so debilitating that she had been telling people not to come by and see the baby. Although it seemed things had been smoother after the birth of their first daughter, Emily, about three years earlier, we had all been thinking maybe this was a normal postpartum reaction: a mix of hormones furiously adjusting, lack of sleep, and the distraction of caring for a newborn. This was what the physicians and nurses had suggested in response to Dave's concerns the week before. But suddenly Jen's condition didn't seem normal anymore.

My mom and stepfather agreed to take the kids, and as a teenager, my job was to play with Emily and distract her from her mom's absence. What unfolded over the next couple of days happened really, really fast. Jen's headaches and dizziness were attributed to a cyst on her brain, which turned into an emergency surgery in which her skull was opened, which turned into days spent looking pale and small in her hospital bed, her head swathed in white gauze. For the rest of the family, those agonizing days were spent switching between helping care for their little girls and tagging out with Dave to sit by Jen's side as she recovered.

I couldn't help but watch Dave's face as he watched his wife's. They'd always been so in love, so in sync with each other, and that bond between them was palpable even in this horrible moment filled with fear and uncertainty, with sadness and anger that no one knew where to direct. Her eyelids fluttered. His normally steady jaw drooped, as if to draw any pain from Jen's body into his own. His thumb swept the back of her hand lightly, sweetly. When a hank of blond hair slipped over her freckled cheek, Dave tucked it back behind her ear as tenderly as if he were stroking one of their daughters' faces. I remember thinking that the look on Dave's face as he regarded his gravely ill wife was an expression of the intense love that human beings live and die for, and that we should all be so lucky to have someone who looks at us the way Dave looked at Jen. They were together in sickness and in health, until death — years later — ultimately forced them apart.

Life-threatening or unexpected health issues have the potential to redefine a relationship. For some, that can mean starting a new chapter of deeper connection, but for countless others it's the end of everything.[1]

KARMIC INSTABILITY

The first wedding I ever officiated was for two friends from graduate school whom I had introduced at a local bar following a costume-themed house party. Sam and Alex had instant chemistry on the dance floor that night and started dating that week. Fewer than three months into this new relationship, disaster struck in the form of Tropical Storm Lee, which crashed through the town we lived in and flooded Alex's apartment building; Sam invited Alex to move in while the cleanup crews did their work, but temporary turned permanent, and a few years later, from their new shared nest in a new town, they asked me to perform the wedding ceremony. I was delighted and honored.

Sitting down to write the words for their wedding ceremony was the first time I had thought critically about that common nuptial phrase "in sickness and in health." I began to wonder whether the phrase really has any teeth to it, and whether the concept had been well studied. How much do we really know about whether people actually *do* stay through the "in sickness" part?

Jen was sick on and off for years. What the doctors had initially thought was a cyst turned out to be brain cancer, an astrocytoma. There would be days, even months, when it seemed the cancer was behind her, and then days and weeks when it was an effort for her to even leave the house. Then there were the horrendous brain surgeries every few years, like clockwork. From the moment she was diagnosed, the family unit had reoriented itself to revolve almost entirely around her illness. In addition to the stress of never knowing when — or *if* — she would get better, Jen and Dave faced enormous expenses, including traveling internationally to try nontraditional treatments, amassing a mountain of medical bills despite their good insurance.[2] As a family, we rallied around them, offering emotional support and pitching in financially when things got tight. Friends and family spoke of their admiration for Jen's fortitude, but I couldn't help noticing that people also murmured their admiration for *Dave's* fortitude. I would later learn, and understand why, some clinicians and researchers call some cancers a "we-disease."[3]

Jen and Dave faced all the predictable challenges and struggles that come with marriage and raising children, along with the extra burdens of navigating an unpredictable illness. The pressures were enormous, but Dave always showed up, never wavering in his love or readiness to care for his wife and their daughters. At one point, they shared with me that the two of them had talked openly about whether their relationship could weather the storm of the cancer. Each time they got to the point where they began to doubt the strength of their pair bond, they actively chose to stay together; the intimacy they had forged as young lovers had become a sustaining force.

Research has shown that not all intimate relationships are able to weather the same kinds of storms. Studies of couples who have lost a child suggest that extreme trauma and grief tend to either split the parents up or bring them closer together.[4] In one such study, many couples reported a subsequent decline in sexual activity, with some noting it was a painful reminder of how the child they lost had initially been conceived.[5] And in one German study of partners and ex-partners of cancer patients, among those who had split up more than half (nearly six out of ten) reported that cancer had contributed to the separation.[6]

If you're watching the person you love struggle to survive an illness, things aren't going to be the same — for better or for worse. When weathering the storms of illness, grief, and loss, the goal should not be to get back to how things were. Instead, "success" is about managing adversity and embracing a "new normal."

Social psychologists have described a uniquely human phenomenon known as the "just-world fallacy," which is a fundamental belief we have that the universe is fair, and therefore bad things happen for a reason. This is at least part of the reason we have a hard time understanding when someone we perceive as innocent is wronged — and when we encounter circumstances that we have a hard time understanding, the human mind tries to come up with a rationale, which can sometimes lead us to unrealistic conclusions.[7] A man is mugged in the park while jogging; someone asks, "How late at night was he out?" It's fundamentally the wrong question, but our unconscious bias leads us to insist he must have done something to precipitate this horrible and statistically rare event so that we don't have to wrap our minds around the fact that there are people out there who cause violence without provocation.

This bias carries over into relationships. A couple that loses a child, for example, seeks meaning in the meaningless as a way to explain and justify their trauma — and pretty soon they can start to find sources of blame in each other.[8] Sometimes a parent is guilty of actions born of not knowing better or not caring, but more often this type of questioning merely creates a vicious cycle of resentment and blame.

Our intimate relationships provide safety, stability, and reassurance, even and perhaps especially when we are in pain. They are both a container and a salve for our psychic wounds. But in our worst states of vulnerability and anguish, our partners can become a target of our pain; in some studies, couples who experienced the sudden death of a child directed their psychological distress and hostility toward each other, perhaps unintentionally.[9] When we are intimate with someone, we feel safe enough to lash out, with the often unspoken understanding that the ties that bind us to our partner are strong enough to survive our emotional attacks.

But when those ties aren't strong enough, when we begin to damage the very structure that provides us refuge, the pain we feel is compounded. Whether the couple stays together or not, a loss of such magnitude is just too intense a trauma for the relationship to stay the same as it once was.

THE HEALING HIGHWAY OF CARE

My Kinsey Institute colleague Sue Carter and I walked across campus one afternoon, noticing the first signs of spring. Daffodils bloomed through the mulch, birds sang from the branches of newly green trees, and college students traipsed around dressed as if the temperature was much warmer than it actually was. Pausing in front of a row of budding magnolias, Sue noted that one of the big-picture findings of her fifty-year research career could be boiled down to a simple phrase: "Love truly *is* the best medicine." She was referring to the undeniable evidence that healthy and positive close relationships can result in elevated oxytocin and physiological states that actively improve our health; that love and the physiological processes that make such intense feelings and joys of connection possible reduce inflammation, improve immune function, regulate the autonomic nervous system, and may even improve the health of our gut microbiome.[10]

Another series of studies reveals that these interactions between relationships and physical health are bidirectional, meaning that just as relationship dynamics can affect our health, our health can also affect our relationship dynamics. In a pack of studies on how individuals and couples experience and cope with illness led by my collaborator Amanda Gesselman, our own lab has found that people with epilepsy, breast cancer survivors and their spouses, those with trauma from adverse childhood experiences, and young men living with HIV all exhibit a similar pattern in their relationships: Their health affects their relationships in measurable ways, and in turn, those relationships affect their health.[11] Sexual function or loss of intimate contact can be a major player. When illness lessens sex drive or makes sex less possible or pleasurable, the partnership suffers.

When illnesses impact sexual and reproductive organs, medical decisions around treatment can place an especially heavy burden on couples' intimate lives. In our work with breast cancer survivors, we observed a consistent finding that many women who undergo cancer-removing mastectomy (the partial or complete surgical removal of one or both breasts) struggle to feel comfortable in their new bodies, and often feel embarrassed by surgical scars.[12] This inevitably impacts the way they behave when trying to be physically and emotionally intimate with their partners, and in some cases includes a lack of desire to be intimate at all. Survivors report grappling with any number of complicated emotions, from shame, anger, and sadness to a sense of mourning that a part of themselves that has defined their femininity has been altered or lost (this is one reason many women's health advocates have argued for breast reconstruction surgeries to be covered by healthcare insurance, as studies have shown this can dramatically help survivors with their psychosocial health).[13]

The partners of women who have had mastectomies, too, face emotional obstacles when engaging in sexual intimacy — the uncertainty of whether they should acknowledge or ignore their partner's scars; picking up on their partner's discomfort but not knowing what to do or how to act; the pain of having to confront visual reminders of their loved one's trauma.[14] One man I spoke with told me about how once, while they were having sex, he had traced a finger along the twin scars of his wife's mastectomy, feeling a sense of awe at her bravery and resilience. "No," she had said simply, grabbing his hand, and he understood. While *he* saw his wife's scars as a source of her strength, for *her* they triggered painful memories of grueling treatments and the fear of leaving her husband a widower, raising their young children alone.

The many medications that go hand in hand with illness can affect aspects of our sex lives as well. Some hormonal drugs, for example, can affect libido; some cardiovascular medications can reduce sexual function (penile erection and vaginal lubrication); other drugs might even cause romantic desires and feelings to be altered.[15] Some psychiatric drugs can have similar effects. In one of our unpublished studies, a startling 11 percent of adults using antidepressant medications reported experiencing blunted romantic feelings as a side effect, consistent with a neuropsychiatric theory on the impacts of regulating serotonin proposed by Helen Fisher and J. Andy Thompson.[16]

One study from Hong Kong of female breast cancer patients and their husbands found that marital adjustment to the realities of breast cancer was strongly associated with the patient's outlook on the world, their relationship, their treatment, and even their marriage.[17] In other words, our relationships and our health go hand in hand. These are the complex emotional dances couples navigate, in the best cases trying to support each other as needs change in the wake of health concerns, illness, and survivorship.

When one partner suffers a trauma, sometimes we hear things like "Well, s/he's the one who's actually going through it." However, while it's true that others often can't fully relate to the personal battle of fighting illness, when it comes to a healthy relationship this statement isn't entirely true. Even if we aren't the person with the main burden of a disease or illness, we are affected. The literature on caretaking provides mountains of evidence of the toll managing a family member's illness — termed "caregiver burden" — can take on partners, parents, kids, and the self.[18] Illness damages our bodies, but it can also damage our intimate relationships, *precisely* at the moment we need them the most.

Jen battled her brain cancer for thirteen years. There were days she seemed to be in the clear and days she seemed almost gone; sometimes we held our breath and sometimes we let it out. Five craniotomies. Countless treatments, several experimental. After she had bounced back so many times, we perhaps forgot that she was mortal after all.

When it became clear that her illness had gained too much ground in the ongoing war, I was deep into my own career some 800 miles away. But even from this distance, I noticed that over the course of Jen's illness, Dave had adapted in ways that were hard to fathom. His whole life — career, socialization, ambitions — had shifted to accommodate his wife, her needs, and the needs of their family, which by that point included two beautiful and remarkably resilient little girls, two dogs, and a turtle named Tommy (which we only much later learned was a female). While Jen was the one fighting cancer, both she and Dave had been out on the battlefield, and they would rise and fall together, time and time again.

Jen had always been my hero. But the sicker she got, the more I realized that Dave was my hero, too. Their story is a remarkable example of the resilient power of intimate love between two human beings.

IN SICKNESS AND IN HEALTH

Part of our legacy as the intimate animal is our capacity to bond with others — not just with a romantic partner but also with family members, colleagues, and friends.[19] And in times of illness or trauma, these social relationships can become an essential source of support and care.

Work by psychologist Julianne Holt-Lunstad and others has demonstrated that the absence of quality relationships is a strong predictor of disease and mortality.[20] In one large meta-analysis comparing data across dozens of studies with a total of over 300,000 people, researchers found a 50 percent increase in survival likelihood as a result of stronger social relationships.[21] When scientists control for the quality of a relationship, even an unsatisfying one might be more beneficial than no relationship at all when it comes to health outcomes. It's not unlike how having a helicopter parent anxiously hovering around a child's every move is not ideal for his or her development, yet it's still better than having no parent or a disengaged one — even if the overly involved kind adds a different kind of stress.[22] The body responds and adapts to social support in whatever form it comes.

There is an interesting nuance to the research on spousal support and social support more generally. While having a supportive partner and close relationships is important, especially in times of urgency or crisis, it's the *perception* of support that matters most.[23] That is, it's not whether you are always available, but whether I believe you'll be there when I need you most.

From the moment she was diagnosed, our whole family became a sort of galaxy orbiting around the black hole of Jen's health. Her long battle for survival gave us something to rally around, something to unite us in the face of the unimaginable. When she died, we all remained extremely close, but without a center our universe faltered.

In the wake of that loss, we did the only thing we could. We turned to Lexi and Emily, the daughters Jen had loved so fiercely but would never get to see graduate from high school and then college, or watch as they got married. We are still a very close family, in many ways still orbiting around Jen's ghost.

Some evolutionists have argued that caring is *the* trait that ultimately facilitated humans having such remarkable success as a species in terms of survival, reproduction, and global distribution. In particular, sharing — first of food and childcare, later of feelings and intentions — allowed everyone to pitch in and provide for the group.[24] With everyone working together, there was additional high-nutrient food, allowing humans to evolve larger brains that facilitated more complex social interaction, and eventually more complex social structures. Sharing, as an outgrowth of caring, allowed us to be a superorganism, a collective that helped its members and made sure everyone was cared for. In other words, care is not only at the center of our romantic relationships but may have been the key to our species's evolutionary success.

Over the last decade, scientists have paid increasing attention to the remarkable power of caring, including its evolutionary foundations and significance to our health, well-being, and ability to thrive. While many people are familiar with the fight-or-flight response, the mobilizing of a physiological and behavioral impulse in the face of danger, it turns out this may be a more commonly male response to threat. Females may be more likely to respond with what biopsychologist Shelly Taylor calls tend-and-befriend: the attempt to protect offspring (tend) or seek out a social group for mutual defense (befriend).[25]

It's easy to assume that traits such as caring, cooperation, empathy, and fairness are uniquely human. But decades of work by primatologist Frans de Waal shows us that other primates and mammals sometimes exhibit similar patterns, and that these social tendencies are critical to their survival.[26]

Other research reveals that humans are not the only animals with the capacity for grief or extreme empathy for others in distress.[27] In 2016, a family of elephants in Samburu National Reserve in northern Kenya was filmed continuing to visit the body of their matriarch for weeks after her death, circling her skeletal remains, probing them gently with their trunks as if they were still trying to process their enormous loss.[28] Magpies have been observed burying the dead under twigs and grass, and chimpanzees have been known to caress those who are nearing death, mourning and grooming them after they are gone, and avoiding the place where they died out of apparent grief.[29]

Yet the relentless devotion to another's well-being while they are still alive is not something we see in the animal kingdom often. In most species, if a member of a pair bond gets sick, their partner will not stay by their side or display the caring behaviors we observe in human beings. Though many primates will groom each other to remove disease-carrying ticks, lice, and fleas, many species have disease-avoidance adaptations that result in them actively staying away from those that seem ill to prevent the spread of a potentially infectious disease.[30]

In the larger natural world, the fact that an animal will cut bait and run when disease strikes their kin is preordained — and if we follow evolutionary logic, we might expect humans to do the same. Rational choice theory, or the idea that humans are rational beings who exercise free will and are in control of their decisions, would likewise expect humans to avoid risks — in this case, the risk of contracting an illness — that are not in their self-interest. Similarly, if we assume that the purpose of romantic love is to ensure the transmission of our genes via reproduction, we might expect humans to abandon a partner who is too sick to conceive or may not live long enough to have children. This behavior would be consistent with a phenomenon biologists sometimes call the Concorde fallacy (named after the Concorde, a British-French supersonic passenger plane whose production costs famously far exceeded the original budget), which holds that individuals shouldn't continue to invest in a sunk cost and should instead reallocate resources toward new opportunities.[31] But in matters of human love, the rules don't always apply.

It is precisely because of the intense power of romantic bonds that we do things for our partners and for our relationships that are unlike what theories of the natural world and other behavioral science models would typically predict. In other words, it is a consequence of our intimate instincts that humans don't abandon their pair bonds lightly.

Our commitments to our loved ones, through thick and thin, are truly remarkable. To live and die for love is not what makes us human per se, but it is one of the most striking hallmarks of our humanity. A philosopher friend has teasingly asked me on more than one occasion: "Do you think we are human because we care so much, or do we care so much because we are human?"

My answer is always consistent: both.

THE POWER OF CARE

Now more than ever — as we find ourselves navigating the dark, open, and uncharted waters of a human intimacy crisis, rising global environmental threats, pandemic instability, and an increasingly polarized political climate — understanding the science and power of caring may very well be critical to our own species's survival. In the face of these existential threats to human health and well-being, our thresholds for caring may ultimately define the next chapters of our relationships and our evolutionary legacy on this planet.

Humans do something remarkable in our close relationships, which is to spell out explicitly the expectation that we will care for each other. We want someone in the trenches with us, someone who will tamp down the entirely natural urge to flee the scene when illness or adversity extracts its financial and psychological toll, and turn toward us instead. The natural world is unpredictable, and part of what we gain in the pair-bonding deal is a partner to help us respond creatively to modern challenges, even if we no longer need to be kept on our toes about changes in food sources or predator threats.

These intimate partnerships do more than simply help us withstand the vicissitudes of life: They help us heal from trauma, recover from illness, and maintain our mental health. That's why I have long believed that in addition to prescribing medication and suggesting treatments, physicians of all kinds would do well to prescribe "relationship medicine": tools that patients can use to maintain the health of their partnerships during times of illness. It might sound like odd advice coming from an oncologist, a cardiologist, or just your general practitioner, but I believe that investing in your relationship when you are ill or unwell is one of the best things you can do for your health. Intimacy helps to heal us and make us whole.

That's why it's especially painful when an intimate connection is severed in times of illness. And yet, the capacity of humans to endure emotional hardship together is an important part of what makes us an intimate animal.

Chapter Ten

LOVE AGAIN

My research partner and I made it about sixty seconds in the heat waiting to be picked up outside our hotel before we were both coated in our own sweat. The humidity was extraordinary. Amanda and I had come to Thailand for an international conference on gender and sexuality, but we had carved out two days for this side trip to the jungles of Khao Yai National Park, where we intended to witness a powerful lesson about pair bonding while fulfilling one of my lifelong nature-expedition goals.

Our guide, Tata, arrived in a steel-gray pickup truck with bench seating installed in the bed. A young but seasoned guide with years of experience escorting visitors through Khao Yai, the third-largest nature preserve in Thailand, Tata took one look at me, shook his head, and sent me back into my bungalow to change; after more than ten years in the lab, I had forgotten that tromping through a jungle in shorts maybe isn't a good idea. Once I was properly pantsed, Amanda and I piled into the back of the truck and it set off toward the heart of the park.

The open-air nature of our ride allowed us to take in the heady smells of lemongrass and earth as we absorbed the music of the jungle: chirping cicadas, choruses of birdsong, wind whispering in the grasses and trees, and the steady hum of distant rushing water — a cacophony of sounds that reminded me of an orchestra tuning up before a performance. We marveled at the dense greenness of everything; we would experience the reason for all this lushness a bit later in the day when the sky opened up for an hour of rain, as it does most days in the region that time of year. All around us, Tata pointed out a celebration of biodiversity: monitor lizards, scorpions, a tiny species of deer that communicates in grunts. We heard and followed the guttural bark-like call of the great hornbill, and with our binoculars watched a pair shake their enormous multicolored bills, unfold their five-foot-plus wingspan, and fly from the trees above us.

But our eyes were peeled for one animal in particular: gibbons.

Gibbons are small apes that mainly live in the jungles of Southeast Asia.[1] They are relatively close cousins to humans, despite being known as "lesser apes" rather than "great apes" like humans, chimpanzees, bonobos, gorillas, and orangutans. While about 15 percent of all primate species engage in social monogamy, gibbons are the only apes other than humans that exhibit patterns of pair bonding, and those patterns are not dissimilar to those of humans. They also serve as a wonderful example of how romance looks in the natural world. You see, when two gibbons are intensely bonded to each other, they don't just mate and engage in mate guarding and territorial defense. They also "sing" to each other.[2] Some primatologists believe that gibbons will only sing one particular song to one particular partner, ever. That is to say, if a gibbon loses a mate, it will not sing the same way to another for the rest of its life. Humans, metaphorically speaking, do the same thing. No two loves are alike, and in those instances when we try to sing the same old love song to a new partner, we're setting ourselves up for disaster.

As much as we may want a love to last until death do us part, things don't always work out that way. We leave or we are left, or we break up simply because we realize it's not working and decide to go our separate ways. People get sick, really sick, sometimes to the point where they're no longer themselves, as with Alzheimer's disease or dementia. And eventually everyone dies, sometimes tragically and sooner than we expect. In the aftermath of love lost, we may feel like we'll never sing again. But while we might never sing the same song, we can, in time, learn a new melody.

LOVE AGAIN...AND AGAIN...AND AGAIN

Despite what Disney princess movies and every rom-com ever made would have us believe, there is no such thing as a "soulmate." In fact, statistically speaking, on a planet with a population of nearly 8 billion human beings, there could be thousands and thousands of people out there with whom we could have meaningful, exciting, and intimate relationships over the course of a lifetime.

The idea that there is one and only one person out there who "completes" us is romantic and alluring, but it's also a fantasy, and one that often keeps us stuck after a relationship ends, especially a long-term one. We are not born to fulfill one particular preordained romantic destiny; the human heart is so much more complex and resilient than that.

But even if we don't believe in soulmates, it can be hard not to get hung up on "the one that got away." I was reminded of this recently when a colleague described running into an old childhood friend whom he considered a sort of missed connection in his life. They had always cared for each other, and they shared an underlying flirtation, a palpable enough attraction. But...the circumstances were never quite right. They were rarely single at the same time, and in a few instances when they were, both were bound up in their own lives. She was married now, with a baby on the way, and he insisted that he couldn't be happier for her, but there was a wistfulness in his tone as he told me the story that suggested a sense of longing for what could have been.

My friend could imagine an alternative version of his life with this person as his partner. Most of us know the feeling; it's easy to romanticize "what could have been," even though in general we tend to overestimate alternatives (leading to undue regret).[3] Even if they had gotten together, the reality of that relationship would likely have been far more complicated than what he was envisioning. Even if we have one partner for decades, we inhabit multiple relationships with them, as we, our partner, and our partnership change over time. Psychotherapist Esther Perel makes this point well.[4] As she puts it, "Most of us are going to have two or three marriages or committed relationships. Some of us will have them with the same people."

Publicly available marital data from around the world tell us that many couples who have spent a significant portion of their lives in a committed pair bond will at some point find themselves facing the prospect that their relationship has run its course.[5] One report noted that the average length of marriage in Rome was roughly eighteen years, with a close to 30 percent divorce rate; in Paris it was roughly thirteen years with a more than 50 percent divorce rate; and in Mexico City the average was twelve years with a much lower 15 percent divorce rate.[6] Another report shows that in the United States, divorce rates have somewhat declined in recent years, with 80 percent of first marriages formed in the 2010s still together in 2025.[7] Of course, these are averages, but what these numbers tell us is that while many couples are together for a substantial duration of their lives, not every relationship remains intact until death do us part. And since we know that people change over time, we can expect both ourselves *and* our partners to change over the course of a relationship, sometimes together, though sometimes apart.

In 2007, Gabriele Pauli, a maverick German politician, stirred up quite a bit of controversy within her male-dominated, mainly Catholic party by suggesting that all marriages should have a seven-year expiration date — at which point both partners would have to say yes to another round or else the marital union (at least the legal aspect) would simply dissolve. Time-limited marriages aren't entirely unheard of in the world, but the concept certainly challenges conventional views toward the lifelong commitment expectations of marriage.[8] One woman I spoke with, who recently lost her husband of more than forty years to complications from Alzheimer's, was a committed Buddhist when she agreed to marry him; to satisfy her valuation of the constancy of change, the two agreed to check in every five years and make sure they both wanted to continue being married to each other. And check in they did, eight times in total over the course of their long, changing, and ultimately happy and satisfying marriage. They never formally renewed their vows, but they did make a conscious decision about whether the people they were at each five-year milestone still wanted to walk the path of life beside the other.[9]

We would do well to make conscious decision-making like this a staple of our intimate lives. Most relationships are a mixed bag of things we like and don't like, and hopefully there are more positives than negatives. After a relationship ends, instead of dragging that past relationship baggage into the next, we can unpack what worked and what didn't and use those learnings to better understand what we want and need from a partner. If you realize that trust was an issue in your last relationship, or that the physical chemistry was there, but you drifted apart because you had so few shared interests, or that you were drawn to a partner who proved to be emotionally unavailable, you may uncover a pattern you've been subconsciously repeating. Conversely, we can also identify the things we want to replicate — and while we may not find someone who meets all of our criteria, the better we understand what we're looking for, the more likely we are to find it.

LOVE IN OUR GOLDEN YEARS

The popularity of OurTime, a dating platform intended specifically for singles fifty years and older that was for a time the fastest-growing dating site in North America, is proof that it is never too late for love. As we get older, especially over the age of sixty, our bodies, priorities, and lifestyles continue to change in ways that shape our intimate lives. But the pursuit of romantic love and the intersection of love with sex and other forms of intimacy don't die when we age, even if those pursuits no longer serve the biological imperative of reproduction.

In fact, research shows that older adults, including those in retirement communities, are having a considerable amount of sex and using sex-enhancing pharmaceuticals like Viagra and Cialis.[10] These older adults are also among the most vulnerable to sexually transmitted infections, with particularly troubling rates among older adults (sixty-five and over).[11] This is due in part to a long history of poor sex education in the United States and other countries, which is further compounded by sex-negative attitudes and generational stigmas around talking about sex for pleasure.

We do see that as people age there is a general decrease in sexual frequency. As our bodies become more worn, we experience various hormonal changes, often combined with the medications many take for chronic health conditions that affect libido or even the ability to have sex.[12] One national study led by gynecologist Stacy Lindau of over 3,000 adults between the ages of fifty-seven and eighty-five showed that the number of people engaging in sexual activity within the last year declined with advancing age, with women being less likely than men to report sexual activity, and that roughly half of sexually active men and women reported at least one bothersome sexual problem, including low sexual desire (43 percent of women), difficulty with vaginal lubrication (39 percent of women), inability to climax (34 percent of women), and erectile difficulties (37 percent of men). When controlling for factors like overall health and medication use, however, the age-related declines in sexual frequency appear to not be quite as steep — this is likely because many medications taken in later life, particularly cardiovascular and psychiatric drugs, have known sexual side effects.[13]

Some of my own collaborative studies have confirmed and expanded these age-related patterns in sexual expression. In one study with sociologist Lisa Miller, we found that men experience declines in frequency of sexual thoughts and sexual activity later in life than women, creating what has been called the gendered "double standard of aging."[14]

But it isn't only that our bodies are aging. We also tend to have sex less often as our intimate relationships age, as passionate love matures into a love that is more comfortable and companionate.[15] These three interrelated factors affecting our sexual lives — biological age, medications, and relationship age — are often treated as part of the same "problem," when in fact they require different approaches. If we want to have more sexual activity in our seventies, it isn't enough to simply join a gym and avoid medications that decrease libido. We also have to make sure we invest in our relationship and maintain a clear-eyed understanding of how our intimate bond with our partner has changed over the decades.

It is very possible to have a healthy sexual life decades past our reproductive prime. Research into geriatric populations bears this out. Though older adults may be having less sex, at least some of them appear to be having *better* sex.[16] In a separate study with Peter Gray and Amanda Gesselman, we found that sexual satisfaction increased with age in women, though not in men. This may be because older women are more comfortable with their bodies, because they know what they want, and because they are more assertive with sexual partners.[17] It may also be that, because of menopause-associated bodily changes, older women are engaging in behaviors that are more conducive to female sexual satisfaction, like slower sex, more foreplay, receiving oral sex, and use of lubricants.

These findings show that when health issues are not a barrier, many people still have active romantic and sexual lives well into older age, whether that's with a single pair-bonded partner over a lifetime or with new partners as they cycle in and out of relationships over the course of their adult lives.

While the desire for emotional and physical intimacy may never go away, our findings also show that what older people prioritize in their intimate lives is different. For instance, our Singles in America study found older adults are the least likely to "settle" when dating; they are more likely to know what they want, and they aren't rushing against a biological clock or other social pressures. Older adults are also positioned to focus more on what they've learned from previous relationship experiences and on what they want now — and are able to experience deeper, more satisfying connections as a result. As one seventy-seven-year-old woman told me, she and her boyfriend enthusiastically have sex every day, sometimes even stopping for a brief conversation or a laugh in the middle of the act — they prioritize and enjoy their sex life together. We have a lot to learn about love and sex from our elders.

Considerations about relationships and sexual expression later in life are culturally variable. Not all peoples or cultures view sexual activity as something done for the purpose of pleasure to draw a couple together, or even as something appropriate in later life.[18] But many acknowledge the enduring bonds of companionship, which can grow stronger as we age and see our partners in the light of different life experiences we've shared (they were a good dad or a terrific grandma, or they had a dedicated career). As a couple from Argentina who had married in their sixties once told me, after having both been previously married and raised families, their union had little to do with lust but focused instead on companionship, enjoying the world together while enjoying each other. There is a term for this in Mandarin Chinese, *zhiji*, which refers to a relationship in which someone can know you as well as you know yourself: a sort of intentional soulmate, with whom you needn't be sexual or even necessarily romantic.

A senior colleague of mine who is in her seventies recently celebrated her fiftieth wedding anniversary. Over the course of their marriage, she and her husband raised two sons while juggling the demands of their ambitious academic careers. When I asked her how they made it work for so long, she replied, "Oh, there were definitely times when it would have been easier to walk away. But we were committed to each other, to the relationship, and to the life that we were building together."

Change can divide us or bring us together. Neighbors recently sent their youngest off to college, and this couple has found new ways to reinforce their bond despite their so-called empty nest, including taking a walk together with their poodles every evening and forming a routine that encourages connection via the nurturing of two very well-behaved albeit spoiled dogs.

Love also comes with a suite of different issues later in life. Sometimes they relate to aging bodies, different resource needs, or new priorities. We also experience psychological changes as we age that can impact our relationships: Studies suggest that people become more altruistic and generous as they age, which can in turn influence behaviors and expectations in couples.[19]

Aging can also pose challenges to dating in midlife.[20] Middle-aged, educated women have some of the greatest trouble on the mating market, as evidenced by a complaint from a forty-seven-year-old professor I know who kept getting responses on dating apps exclusively from much older men. "Are the only guys I can date sixty-five years old?" she asked me frustratedly, as prelude to a discussion about loosening her stated preferences on the app and widening the range on the variables she was searching for.

Age discrepancies aren't the only issue. In heterosexual relationships, men tend to be relatively older than their partners, a pattern observed cross-culturally.[21] However, as both partners age, discrepancies in age tend to become more pronounced. One woman I talked to confided to me that when she was younger the eighteen-year age gap with her partner hadn't seemed to matter, but now that she is in her sixties and he in his eighties, the relationship required more caretaking and less romantic intimacy than she had ever previously considered. While it does make evolutionary sense that on average women tend to prefer somewhat older men (this is consistent with preferences for mates with greater resources), based on the mere fact that women tend to live longer than men it might make more sense for women to date younger men and reduce the risk of spending their late-life years on their own after the death of a partner.[22]

The presence of children, even grown ones, influences our intimate lives in myriad ways, and as we age, the likelihood of having children and grandchildren increases. While young children impact our intimate lives by limiting our available energy and physiological attention, adult children may have any number of reasons for disapproving of their parent's or grandparent's new partner, or may even be uncomfortable with the mere fact of their parent's or grandparent's romantic and/or sexual life.

For example, I spoke to one middle-aged man and his husband, who were worried the younger woman one partner's elderly father was dating was trying to take financial advantage of him; the son and his husband did not support them getting married without a rather restrictive prenuptial agreement, which in turn upset their stepmother-to-be, who noted she would be the one assuming responsibility for their father's care as he aged. Beyond practical concerns about finances or caregiving, however, adult children often resent the presence of a parent's (or grandparent's) new partner on a more emotional level. It seems that many can never quite outgrow the instinctive distrust or apprehension about a new person stepping into a role formerly occupied by someone they consider a parent.

But as we — and our parents — age, we inevitably gain perspective about how intimacy changes through time. Our intimate lives are a marathon, not a sprint. Talk to your parents and grandparents. What concessions are they happy they made? What regrets do they have about things they never did? The answers to these questions become data points we can look to as we navigate our own relationships. When we open a dialogue with those who have run the marathon, we can learn from their experience.

Most important is to remember that as we age, our priorities change, and our sexual and romantic lives change with us. And the better we pay attention to our shifting desires, the evolving needs of our partners, and the intimate contexts in which we operate, the better we can navigate later-in-life love.

A NEW KIND OF LOVE SONG

About thirty minutes into our ride through Khao Yai National Park, Tata pulled off the trail and put his finger to his lips.

"Hear that?" he asked.

We strained to listen and could just make out a new sound tuning up among the jungle orchestra, coming from a heavily wooded area just below us.

"We're going down there," Tata said.

We climbed out of the truck, tied up our leech socks, grabbed our binoculars and cameras, and headed toward the trees. With our hands raised over our heads, we marched through weedy grasses so tall they came up to chest height on me (and I'm over six feet). At one point, Tata pointed toward a pile of dung and identified it as belonging to an elephant that had passed through this area as recently as the day before.

We soon plunged into the shaded coolness of the tree canopy. A few steps in, Tata put his finger to his lips again and pointed upward. Amanda and I followed his gaze, and then we saw them. One, then two, then three, in the branches just above us: a whole family of white-handed gibbons. We saw a mother carrying a baby and a juvenile bouncing athletically from branch to branch and tree to tree with the ease and showmanship of a performer in Cirque du Soleil. A few trees over, the male of the family unit sat on a high branch surveying the activity of the jungle, ready to spring into action at the first sign of a threat to his family.

And they were singing. Loudly. Their voices echoed through the trees. It wasn't a call-and-response type of utterance, and it wasn't always clear who was singing to whom, but the effect was eerily beautiful. The breeding pair was chorusing to the morning in what is sometimes referred to as the "great call." We listened and watched for about ninety minutes, completely captivated.[23]

As I stood there mesmerized by this strange and wonderful display, I was struck by the thought that this gibbon pair and their family would experience so much love in their lifetime. As they aged together, they might fend off competitors vying for their territory, or neighbors trying to mate with one or the other. They might produce more offspring, some of which would live and leave the nest, some of which might fall from the trees and die or succumb to predators or pathogens. One might even survive the other and form a new "marriage."

And, through all of it, this family of gibbons would be a unit with a shared bond — and a love song for each other. Not so different, really, from us humans. Watching them go about their business laid bare to me the majesty of the evolutionary process at play in this ecosystem so densely packed with different species, all fighting for survival and reproduction, and some fighting for something akin to love.

A year or so after Jen died, Dave and I went out to lunch. Over Cobb salads, we talked about our lives, career plans, and the vibrant branches of our family, including his and Jen's two fantastic girls. Lexi, then fourteen, was a freshman in high school, and Emily, seventeen, was a senior and deep into the college application process.

Eventually the conversation turned to dating. We chatted about my love life, then we turned to his.

"It's only been a year," he said, sighing, "but people really want me to date. They keep offering — threatening — to fix me up."

Well-meaning friends and colleagues, who recognized Dave as a great partner and provider, thought a companion would be good for him, and that perhaps the girls would benefit from a stepmother figure in their lives. At the very least, they hoped Dave would work through his grief more expeditiously if he had a new crush.

But he wasn't ready.

"They just don't understand. I'm never going to meet anyone like Jen again."

My heart ached in that moment, partly for Dave's loss but also for my own still-fresh sadness over Jen being gone from my life, from our lives, too soon. I understood where he was coming from, and it's a common sentiment among all of us who have lost love. But I also knew from my work and research that it's a profoundly misdirected one.

"You're right," I said. "You'll never meet anyone like Jen."

Dave's face betrayed a bit of confusion — these seemed like harsh words from someone who studies relationships. But I had more to say.

"You'll never meet another Jen. But when you're ready to date, you'll find someone who does something *different* for you. There's no replacement for her. If you decide to pursue a new relationship, it will be something and someone completely new."

I wasn't sure whether Dave would ever even want to find another person to share his life with. But I am sure of the research, which tells us it *is* possible to love again, when we are ready, and if that's what we want. Dave will likely find someone different from Jen and have a relationship that, like the gibbons' song, is quite different from the one they shared. Like so many of us who want, find, and then commit to an intimate connection, only to experience the pain of love lost, he will somehow find the resilience to love again.

That doesn't mean we should forget our pasts; we can still enjoy the memories and how they made us the person we are. For Dave, there's an awful lot to be said for having shared a life with Jen, a woman he loved, and still loves, starting from before the day I met him at my carousel kindergarten birthday party to the moment she let go of her life. Their intimate bond shaped his past, and even with her gone from the world it will continue to shape his future.

Intimacy, by its very definition, is a state we can experience only in the context of another person; it's that magical mix of closeness, trust, and vulnerability that ripens over time and anchors us in the world. It's perhaps for this reason above all that we have evolved as the intimate animal — to allow us as a species to weather the uncertainty of ever-changing physical and social environments. When that bond is broken, we can love again; sometimes better, but always different, because relationships are dynamic, and we learn from the past and adapt to experience.

When we put all the pieces together, the research is clear: It's possible to fall out of love, it's okay to grieve that loss, and it's in our nature as the intimate animal to learn a new song and love again.

CONCLUSION

LIVING AND DYING FOR LOVE

In March 2020, a week before most of the United States, and the world, went into COVID lockdowns, I flew to New York to give a talk at the Matchmaking Institute's annual Global Love Conference, where I presented my latest research on dating and gender and the #MeToo movement. At the reception later that evening, held at a penthouse in a Hudson Yards high-rise overlooking the river in Midtown Manhattan, conversations between the gathered participants were muted. Everyone present was in the business of love, of connection; yet we sensed the world was about to experience a seismic shift, the aftershocks of which we would feel for years, perhaps decades, to come. People were anxious. No one shook hands or hugged, and being at arm's length from one another was unfamiliar to this crowd of researchers and matchmakers who spend our lives digging around in the trenches of human connection and intimacy.

By the very next day, the writing was on the wall about the spread of COVID, and my mind turned to a different set of intimate relationships: my parents, who were ensconced in their apartment not far from what would quickly become the epicenter in the United States of what was already now clearly a global pandemic. We did the mental calculus and all agreed that they would ride out the pandemic with me in Indiana. I canceled my plane ticket, packed up their Jeep, and prepared for a long car ride.

Our route brought us past Forest Park in Queens, where almost thirty years earlier I had met Dave for the first time. "When you were a little boy you used to love this park," Mom commented as we turned onto Woodhaven Boulevard. "Do you remember when we had your birthday party at the carousel?"

I did remember.

I hadn't planned on taking any detours, but I found myself turning into the park and a few minutes later we pulled up in front of the carousel. As I stood watching the ring of beautifully painted wooden horses galloping in an infinite loop, I thought of the nerdy little boy with an insatiable curiosity about the world around him, and remembered Jen's light and Dave's introduction into our family.

Observing the few people still playing in the park, my eye was drawn to a scene that could have been pulled from my own memory. On a bench beneath the shade of one of the park's hundred-year-old pine trees, a little boy was holding court as a couple in their mid-twenties doted on him, sneaking each other affectionate smiles. I marveled at these people making entirely new memories at a time when the world was about to face so much loss and so many uncertainties about the road ahead.

I often wonder about the future of our intimate lives. How will climate change shape our intimate relationships, particularly when it comes to people's decisions about whether to have children? How are our intimate partnerships adapting in response to increased global political conflict and economic instability? How will our intimate lives change in a world where technology increasingly mediates our personal and romantic lives?

Technology in particular prompts many questions about the fundamental nature of intimacy. As we do more and more of our dating online, and as artificial intelligence becomes more and more sophisticated, it isn't hard to imagine a future in which advanced chatbots function as romantic surrogates in online spaces; in fact, this future is coming for us already, with Google Trends data revealing a 2,400 percent increase in searches for "AI girlfriend" in 2023.[1] We might feel a sense of intimacy while baring our soul to a doting chatbot, but is this truly intimacy? Can the illusion of human connection ever hold a candle to the real thing? And as the AI bots learn to communicate in ever more human-like ways, will we be able to tell the difference?

In one of our national studies, one in four single adults said they would have sex with a robot if given the opportunity. Yet, nearly half of singles would consider it cheating if their partner did so. When I saw this in the data, my first thought was, *This is a species in crisis.* But the more I thought about it, the more I came to understand that we're actually a species on the precipice of an evolutionary *revolution*. The technological and social world humans evolved in is going through rapid, unprecedented change, and our romantic and sexual lives are along for the ride.

It's exciting and beautiful and scary and worrying to watch and document. And although I do not have all the answers, I do believe that there is something unique to the human animal and the physical relationships we enter into with each other — something that will survive, even thrive, in an increasingly digital future.

I've argued that humans have evolved with an intimacy instinct; that we seek out romantic love not just for reproduction and survival but also in the pursuit of self-expansion. We are a variable and adaptive animal, and so we find ways to make intimacy work, but in the face of a loneliness epidemic and the increasing digitization of our social worlds, our ability to satisfy our drive for intimacy across our lifespan is being challenged more than ever. Failure to appreciate the science behind our choices is something we do at our own peril. Ultimately, if we fail to understand why we do what we do, we will be unable to achieve happiness and satisfaction with our relationships and our lives.

My research has shown that there is such a thing as an "ideal" pair bond; healthy and happy relationships are about balancing our desires and needs in accordance with our competing drives, and bringing them into harmony. And while it's true that we didn't evolve to be happy, we can be, if we so choose.

And herein lies our most hopeful discovery: *We are not prisoners of our drives for sex or love, or even intimacy.* When we're in the throes of passion — the ecstasy of falling in love, the misery of losing it — it can feel out of our control. But our highly evolved forebrains allow us to make conscious decisions that create our opportunities. Life throws us curveballs, and we can choose whether to greet the unexpected as an adventure or allow it to thwart the bonds we've built. Fortunately, we are an adaptive animal.

More than two years after my parents and I made that detour into Forest Park, Queens, I would find myself in another grove of pines, some 800 miles to the west, thinking once again about the profound nature of the promise we as humans make when we decide to build a life together. Even though it was a cool November day as Michelle and I strolled along the winding paths through the wooded grounds of the Tibetan Buddhist cultural center in Bloomington, Indiana, I was practically sweating through my shirt.

I had been planning this moment for months, timing it to coincide with Michelle's parents' visit from their home in Nanjing, China. Along with my mom and dad, who by then had permanently relocated to Bloomington, her parents had become my co-conspirators. Through a series of texts over WeChat, which thankfully has a translation function, I had managed to convey my love for their daughter, my intention to ask for her hand in marriage, and essentially ask for their blessing (something the gender studies scholar in me balked at, but which I nonetheless felt was important as a gesture of respect). As I dropped to one knee and pulled out the ring box hidden in my jacket pocket, I could tell from the look of surprise on Michelle's face that everyone had kept our secret. Suddenly, I was so overwhelmed with emotion I lost my words. It seems, even after a lifetime of examining the intimate lives of others, I'm still sometimes tongue-tied when it comes to my own.

Eventually, I managed to sputter out a proposal. As Michelle and I kissed for the first time as a betrothed couple, our cheeks wet from tears of overwhelming love, I felt like something had come full circle, both personally and professionally.

This book is an exploration of that circle.

As we've examined in these pages, the human animal needs meaningful relationships. Not simply because they bring us joy, but because the fabric of our lives is dependent on them. Intimacy is one of the essential ingredients in our recipe for life. In the absence of romantic and sexual connection, we may survive, but we cannot flourish without the intimate bonds that tether our lives to other humans.

The more we understand our intersecting biological and cultural systems — the tensions between love and sex, including the ways in which we've evolved to want and need intimacy — the closer we can get to answering the question of *why* we are the way we are. And that is where the power of the intimate animal truly lies: We alone in the animal kingdom have the ability to know ourselves and, in so doing, enjoy healthier and more fulfilling romantic relationships.

Humans, the most intimate animal, will always live and die for love. Understanding *why* gives us the power to find and nurture the loves worth living for.

ACKNOWLEDGMENTS

There are so many people to acknowledge and thank for their role in shaping everything shared in *The Intimate Animal*.

Interdisciplinary research is a team sport — and in satisfying science, just as in matters of love, there are always multiple voices that collectively mold the final product. I'm deeply grateful for the many amazing teammates I have enjoyed throughout my career, colleagues who have contributed to my thinking, collaborative research projects, and the studies featured in this book. I am especially grateful to my many collaborators at the Kinsey Institute at Indiana University, and previously at Binghamton University, and in the larger multidisciplinary fields of sex research and relationship science and evolutionary behavioral sciences, who have sharpened my thinking and tested the limits of ideas shared here. There are so many researchers and professionals I admire and am indebted to, most especially my closest collaborators who are also dear friends and who have been generous with their brilliance and both tolerated and celebrated rolling around in complex ideas and datasets together: Amanda Gesselman (my number one co-conspirator), David Frederick, Lucia O'Sullivan, Kristen Mark, Peter Gray, Dylan Selterman, Leslie Heywood, Ann Merriwether, David Sloan Wilson, Kathryn Coe, and the late great Helen Fisher.

And to the full Kinsey Institute community, thank you for your confidence and support and for continuing to inspire me every day with your innovative and impactful research and insights: Alexa Marcotte, Amanda Gesselman, Amy Moors, Brea Perry, Brenda Weber, C. Sue Carter, Camilla Peterson, Christie Hefner, Christopher

Walling, Cynthia Graham, Deb Tolman, Derek Dixon, Dev Montanez, Ellen Kaufman, Garry Milius, Gregory Lewis, India Thusi, Janae Cummings, Jennifer Barber, Jessica Hille, Judith Allen, Julia Heiman, Justin Lehmiller, Lauren Streicher, Lemuel Watson, Liana Zhou, Maggie Bennet-Brown, Maggie Frey, Marie Metelnick, Melissa Blundell Osorio, Myeshia Price, Peggy Maschino, Rebecca Fasman, Sarah Knott, Shawn Wilson, Simon Dubé, Stephanie Sanders, Stephen Porges, Tania Reynolds, William Yarber, Zoë Peterson, and many others. You have all put more wind in my sails than I can ever articulate.

Thank you to all my partners at Match for fifteen years of incredible and impactful partnership exploring the science of dating and human connection in the digital age. Thank you most especially to Amy Canaday, Lauren DeFord, Alexis Ferraro Luerssen, Eva Gallagher, Max Izenberg, Maggie Gillespie, Michael Kaye, Vidhya Murugesan, and many others over the years who have made this work fun and grounded in current affairs.

I am beyond grateful to the full publishing team at Little, Brown Spark. Most especially my editor, Talia Krohn, for seeing my vision and so thoughtfully partnering with me to honor the nuances as we brought this book to the world.

I have enormous gratitude for the wonderful team at Idea Architects, who are so much more than my literary agents: You are trusted friends and thought partners. Thanks most especially to Doug Abrams for years of friendship, and to Wenonah Hoye for your patience and brilliance as a collaborator who helped me make the science resonate and come alive. And thanks to the entire Idea Architects team for your support and guidance: Elizabeth Wachtel, Sarah Rainone, Jordan Jacks, Jenny Davis, Lara Love Hardin, and Rachel Neumann.

And thank you to my family, to whom this book is dedicated, for your love and unwavering support for me and my work: my wife, Michelle; my parents, Helen and Mike; my family Jen, Dave, Emily, and Alexis. *Thank you* for continuously showing me the awesome power of love in all its dimensions.

And last, a great big thank-you to everyone who shared a piece of their life with us for the studies that inform all aspects of this book. I have such respect for those who have vulnerably shared a part of themselves when they "tell me things," who join on a journey to allow research and science to make our intimate lives more informed and more fulfilling.

I feel extremely fortunate to have benefited from the support, counsel, encouragement, and partnership of so many throughout the journey of writing and sharing *The Intimate Animal*. Thank you to everyone who has helped uncover and celebrate the loving, caring, passionate, and curious evolved animal inside us all.

NOTES

Introduction: The White Whale

1. Peter Gray and I explored some of these themes in our co-authored book, Peter B. Gray and Justin R. Garcia, *Evolution and Human Sexual Behavior* (Cambridge, MA: Harvard University Press, 2013), and in Justin R. Garcia, Michelle J. Escasa-Dorne, Peter B. Gray, and Amanda N. Gesselman, "Individual Differences in Women's Salivary Testosterone and Estradiol Following Sexual Activity in a Nonlaboratory Setting," *International Journal of Sexual Health* 27, no. 4 (2015): 406–17.
2. There are several articles I draw from to make the arguments and definitions of intimacy in the pages to come. These include: Roy F. Baumeister and Mark R. Leary, "The Need to Belong: Desire for Interpersonal Attachments as a Fundamental Human Motivation," *Interpersonal Development* 117, no. 3 (2017): 57–89; Arthur Aron, Elaine Melinat, Elaine N. Aron, Robert D. Vallone, and Renee J. Bator, "The Experimental Generation of Interpersonal Closeness: A Procedure and Some Preliminary Findings," *Personality and Social Psychology Bulletin* 23, no. 4 (1997): 363–77; Gurit E. Birnbaum and Danielle Laser-Brandt, "Gender Differences in the Experience of Heterosexual Intercourse," *Canadian Journal of Human Sexuality* 11, nos. 3–4 (2002): 143–58; Catherine Birnie-Porter and John E. Lydon, "A Prototype Approach to Understanding Sexual Intimacy Through Its Relationship to Intimacy," *Personal Relationships* 20, no. 2 (2013): 236–58; Helen E. Fisher, *Anatomy of Love: A Natural History of Mating, Marriage, and Why We Stray,* rev. ed. (New York: W. W. Norton, 2016).
3. Gray and Garcia, *Evolution and Human Sexual Behavior*; Bobbi S. Low, *Why Sex Matters: A Darwinian Look at Human Behavior,* rev. ed. (Princeton, NJ: Princeton University Press, 2015); Erick Janssen, ed., *The Psychophysiology of Sex* (Bloomington: Indiana University Press, 2007); Todd K. Shackelford and Ranald D. Hansen, eds., *The Evolution of Sexuality* (New York: Springer, 2014); Donald Symons, *The Evolution of Human Sexuality* (New York: Oxford University Press, 1979); David M. Buss, *The Evolution of Desire: Strategies of Human Mating,* rev. ed. (New York: Hachette, 2016); Emily Nagoski, *Come As You Are: The Surprising New Science That Will Transform Your Sex Life* (New York: Simon & Schuster, 2015).
4. Helen E. Fisher, *Why We Love: The Nature and Chemistry of Romantic Love* (New York: Macmillan, 2004).

5. Jonathan Gathorne-Hardy, *Sex the Measure of All Things: A Life of Alfred C. Kinsey* (Bloomington: Indiana University Press, 2000); Wardell B. Pomeroy, *Dr. Kinsey and the Institute for Sex Research* (New York: Harper and Row, 1972).
6. Judith A. Allen, Hallimeda E. Allinson, Andrew Clark-Huckstep, Brandon J. Hill, Stephanie A. Sanders, and Liana Zhou, *The Kinsey Institute: The First Seventy Years* (Bloomington: Indiana University Press, 2017); Donna J. Drucker, "'A Noble Experiment': The Marriage Course at Indiana University, 1938–1940," *Indiana Magazine of History* 103 (2007): 231–64.
7. Alfred C. Kinsey, Wardell B. Pomeroy, and Clyde E. Martin, *Sexual Behavior in the Human Male* (Philadelphia: W. B. Saunders, 1948); Alfred C. Kinsey, Wardell B. Pomeroy, Clyde E. Martin, and Paul H. Gebhard, *Sexual Behavior in the Human Female* (Philadelphia: W. B. Saunders, 1953).
8. Allen et al., *The Kinsey Institute;* Drucker, "'A Noble Experiment.'"
9. Donna J. Drucker, *The Classification of Sex: Alfred Kinsey and the Organization of Knowledge* (Pittsburgh: University of Pittsburgh Press, 2014).
10. Donna J. Drucker, "Marking Sexuality from 0–6: The Kinsey Scale in Online Culture," *Sexuality and Culture* 16 (2012): 241–62; Julia R. Heiman, "Alfred C. Kinsey's Legacy and the Kinsey Institute at Indiana University," in *Routledge Handbook of Sexuality, Gender, Health and Rights,* 2nd ed., ed. Peter Aggleton et al. (New York: Routledge, 2023).
11. David A. Frederick, Janell Lever, Brian J. Gillespie, and Justin R. Garcia, "What Keeps Passion Alive? Sexual Satisfaction Is Associated with Sexual Communication, Mood Setting, Sexual Variety, Oral Sex, Orgasm, and Sex Frequency in a National US Study," *Journal of Sex Research* 54, no. 2 (2017): 186–201; Rhonda N. Balzarini, Chantal Dharma, Amy Muise, and Taylor Kohut, "Eroticism Versus Nurturance," *Social Psychology* 50, no. 3 (2019): 185–200; Bianca P. Acevedo, Arthur Aron, Helen E. Fisher, and Lucy L. Brown, "Neural Correlates of Long-Term Intense Romantic Love," *Social Cognitive and Affective Neuroscience* 7, no. 2 (2012): 145–59; Lucia F. O'Sullivan, Corinne F. Belu, and Justin R. Garcia, "Loving You from Afar: Attraction to Others ('Crushes') Among Adults in Exclusive Relationships, Communication, Perceived Outcomes, and Expectations of Future Intimate Involvement," *Journal of Social and Personal Relationships* 39, no. 2 (2022): 413–34.
12. Sigmund Freud, *Three Contributions to the Theory of Sex* (New York: Nervous and Mental Disease Publishing, 1930).
13. This is from a personal communication with Judith Allen, Distinguished Professor and Walter Professor of History at Indiana University and Kinsey Institute historian. Allen intended to include this in her next book, but she died unexpectedly in summer 2024.
14. Ashley K. Randall and Pamela J. Lannutti, eds., *Experiences of Sexual Minority and Gender Diverse Individuals in Romantic Relationships: Heeding a Global Call* (Cambridge: Cambridge University Press, 2025).
15. Helen E. Fisher, Justin R. Garcia, and Amanda N. Gesselman, "Singles in America: Annual Survey" (Match.com), accessed March 28, 2025, https://www.singlesinamerica.com.

Notes

Chapter One: Need

1. Helen E. Fisher, *Why We Love: The Nature and Chemistry of Romantic Love* (New York: Macmillan, 2004); William Jankowiak, ed., *Romantic Passion: The Universal Experience?* (New York: Columbia University Press, 1995); Ted DiDonato and Brittany Jakubiak, *The Science of Romantic Relationships* (Cambridge: Cambridge University Press, 2023); Robert J. Sternberg and Karin Sternberg, eds., *The New Psychology of Love*, 2nd ed. (Cambridge: Cambridge University Press, 2018).
2. Garth J. O. Fletcher, Jeffry A. Simpson, Lorne Campbell, and Nickola C. Overall, "Pair-Bonding, Romantic Love, and Evolution: The Curious Case of *Homo sapiens*," *Perspectives on Psychological Science* 10, no. 1 (2015): 20–36; Robert J. Quinlan, "Human Pair-Bonds: Evolutionary Functions, Ecological Variation, and Adaptive Development," *Evolutionary Anthropology: Issues, News, and Reviews* 17, no. 5 (2008): 227–38; Bernard Chapais, "Monogamy, Strongly Bonded Groups, and the Evolution of Human Social Structure," *Evolutionary Anthropology: Issues, News, and Reviews* 22, no. 2 (2013): 52–65; C. Sue Carter, "Neuroendocrine Perspectives on Social Attachment and Love," *Psychoneuroendocrinology* 23, no. 8 (1998): 779–818; Patricia Adair Gowaty, "Mating Behaviour," in *The Routledge International Handbook of Comparative Psychology*, ed. Jennifer Vonk and Todd K. Shackelford (New York: Routledge, 2022), 175–87; D. Lukas and T. H. Clutton-Brock, "The Evolution of Social Monogamy in Mammals," *Science* 341, no. 6145 (2013): 526–30; Carel P. van Schaik and Peter M. Kappeler, "The Evolution of Social Monogamy in Primates," in *Monogamy: Mating Strategies and Partnerships in Birds, Humans and Other Mammals,* ed. Ulrich H. Reichard and Christophe Boesch (Cambridge: Cambridge University Press, 2003), 59–80; Helen E. Fisher, "Evolution of Human Serial Pairbonding," *American Journal of Physical Anthropology* 78, no. 3 (1989): 331–54; Helen E. Fisher, *Anatomy of Love: A Natural History of Mating, Marriage, and Why We Stray,* rev. ed. (New York: W. W. Norton, 2016); Peter B. Gray and Justin R. Garcia, *Evolution and Human Sexual Behavior* (Cambridge, MA: Harvard University Press, 2013).
3. Jonathan Haidt, *The Anxious Generation: How the Great Rewiring of Childhood Is Causing an Epidemic of Mental Illness* (New York: Penguin, 2024); Julianne Holt-Lunstad, "Social Connection as a Public Health Issue: The Evidence and a Systemic Framework for Prioritizing the 'Social' in Social Determinants of Health," *Annual Review of Public Health* 43, no. 1 (April 5, 2022): 193–213; Julianne Holt-Lunstad, "Social Connection as a Critical Factor for Mental and Physical Health: Evidence, Trends, Challenges, and Future Implications," *World Psychiatry* 23, no. 3 (2024): 312–32; Jean M. Twenge, *Generation Me: Why Today's Young Americans Are More Confident, Assertive, Entitled—and More Miserable Than Ever Before,* rev. ed. (New York: Simon & Schuster, 2014); Helen E. Fisher and Justin R. Garcia, "Mate Choice in the Digital Age," in *The Handbook of Human Mating,* ed. David M. Buss (New York: Oxford University Press, 2023), 777–95; Treena Orchard, *Sticky, Sexy, Sad: Swipe Culture and the Darker Side of Dating Apps* (Toronto: University of Toronto Press, 2024); Vanessa P. Ta, Amanda N. Gesselman, Blair L. Perry, Helen E. Fisher, and Justin R. Garcia, "Stress of Singlehood: Marital Status, Domain-Specific Stress, and Anxiety in a National U.S. Sample," *Journal of Social and Clinical Psychology* 36, no. 6 (2017): 461–85; Eli J. Finkel, Elaine O. Cheung, Lydia

F. Emery, Kathleen L. Carswell, and Grace M. Larson, "The Suffocation Model: Why Marriage in America Is Becoming an All-or-Nothing Institution," *Current Directions in Psychological Science* 24, no. 3 (2015): 238–44; Richard Fry, "Share of U.S. Adults Living Without a Romantic Partner Has Ticked Down in Recent Years," Pew Research Center, January 8, 2025, https://www.pewresearch.org/short-reads/2025/01/08/share-of-us-adults-living-without-a-romantic-partner-has-ticked-down-in-recent-years/.

4. Bella DePaulo, *Singled Out: How Singles Are Stereotyped, Stigmatized, and Ignored, and Still Live Happily Ever After* (New York: Macmillan, 2007); Anna Brown, "A Profile of Single Americans," Pew Research Center, August 20, 2020, https://www.pewresearch.org/social-trends/2020/08/20/a-profile-of-single-americans/.

5. Pamela J. Smock, "Cohabitation in the United States: An Appraisal of Research Themes, Findings, and Implications," *Annual Review of Sociology* 26, no. 1 (2000): 1–20; Sharon Sassler and Daniel T. Lichter, "Cohabitation and Marriage: Complexity and Diversity in Union-Formation Patterns," *Journal of Marriage and Family* 82, no. 1 (2020): 35–61; Catherine G. Campbell, "Two Decades of Coparenting Research: A Scoping Review," *Marriage and Family Review* 59, no. 6 (2023): 379–411.

6. Juliana Menasce Horowitz, Nikki Graf, and Gretchen Livingston, "Marriage and Cohabitation in the U.S.," Pew Research Center, November 6, 2019, https://www.pewresearch.org/social-trends/2019/11/06/marriage-and-cohabitation-in-the-u-s/.

7. Helen E. Fisher and Justin R. Garcia, "Slow Love: Courtship in the Digital Age," in *The New Psychology of Love*, 2nd ed., ed. Robert J. Sternberg and Karin Sternberg (Cambridge: Cambridge University Press, 2018), 208–22; Fisher and Garcia, "Mate Choice in the Digital Age."

8. Bella DePaulo, "Single and Flourishing: Transcending the Deficit Narratives of Single Life," *Journal of Family Theory and Review* 15, no. 3 (2023): 389–411; Yaara U. Girme, Yuthika Park, and Geoff MacDonald, "Coping or Thriving? Reviewing Intrapersonal, Interpersonal, and Societal Factors Associated with Well-Being in Singlehood from a Within-Group Perspective," *Perspectives on Psychological Science* 18, no. 5 (2023): 1097–120.

9. Roy F. Baumeister and Mark R. Leary, "The Need to Belong: Desire for Interpersonal Attachments as a Fundamental Human Motivation," in *Interpersonal Development,* ed. Nancy Eisenberg (New York: Psychology Press, 2017), 57–89.

10. Justin J. Lehmiller, Justin R. Garcia, Amanda N. Gesselman, and Kristen P. Mark, "Less Sex, but More Sexual Diversity: Changes in Sexual Behavior During the COVID-19 Coronavirus Pandemic," *Leisure Sciences* 43, nos. 1–2 (2020): 295–304.

11. Lehmiller et al., "Less Sex, but More Sexual Diversity"; Rhonda N. Balzarini, Amy Muise, Gabriele Zoppolat, Amanda N. Gesselman, Justin J. Lehmiller, Justin R. Garcia, and Kristen P. Mark, "Sexual Desire in the Time of COVID-19: How COVID-Related Stressors Are Associated with Sexual Desire in Romantic Relationships," *Archives of Sexual Behavior* 51, no. 8 (2022): 3823–38; Diana L. Rodrigues and Rhonda N. Balzarini, "Relationship and Sexual Quality in the Wake of COVID-19: Effects of Individual Regulatory Focus and Shared Concerns over the Pandemic," *European Journal of Investigation in Health, Psychology and Education* 13, no. 2 (2023): 460–71; Karen B. Vanterpool, Heather M. Francis, Kirsten M. Greer, Zoe Moscovici, Cynthia A. Graham, Stephanie A. Sanders, Robin R. Milhausen, and William L. Yarber, "Changes in

Marital Relationships over the Course of the COVID-19 Pandemic," *Family Relations* 74, no. 2 (2025): 642–57; T. Jessica Campbell, Maggie Bennett-Brown, Alexandra S. Marcotte, Ellen M. Kaufman, Zoe Moscovici, Olivia R. Adams, Sydney Lovins, Justin R. Garcia, and Amanda N. Gesselman, "American Singles' Attitudes Toward Future Romantic/Sexual Partners' COVID-19 Vaccination Status: Evidence for Both Vigilance and Indifference in a National Sample," *Sexuality and Culture* 27, no. 5 (2023): 1915–38.

12. Bianca P. Acevedo, Arthur Aron, Helen E. Fisher, and Lucy L. Brown, "Neural Correlates of Long-Term Intense Romantic Love," *Social Cognitive and Affective Neuroscience* 7, no. 2 (2012): 145–59; Bianca P. Acevedo, Michael J. Poulin, Nancy L. Collins, and Lucy L. Brown, "After the Honeymoon: Neural and Genetic Correlates of Romantic Love in Newlywed Marriages," *Frontiers in Psychology* 11 (2020): 634; Bianca P. Acevedo, Michael J. Poulin, and Lucy L. Brown, "Beyond Romance: Neural and Genetic Correlates of Altruism in Pair-Bonds," *Behavioral Neuroscience* 133, no. 1 (2019): 18–31.

13. Helen Fisher, Arthur Aron, and Lucy L. Brown, "Romantic Love: An fMRI Study of a Neural Mechanism for Mate Choice," *Journal of Comparative Neurology* 493, no. 1 (2005): 58–62; Arthur Aron, Helen Fisher, Debra J. Mashek, Greg Strong, Haifeng Li, and Lucy L. Brown, "Reward, Motivation, and Emotion Systems Associated with Early-Stage Intense Romantic Love," *Journal of Neurophysiology* 94, no. 1 (2005): 327–37; Acevedo et al., "Neural Correlates of Long-Term Intense Romantic Love"; Stephanie Ortigue, Francois Bianchi-Demicheli, Nisha Patel, Christine Frum, and James W. Lewis, "Neuroimaging of Love: fMRI Meta-Analysis Evidence Toward New Perspectives in Sexual Medicine," *Journal of Sexual Medicine* 7, no. 11 (2010): 3541–52; Stephanie Cacioppo, Francois Bianchi-Demicheli, Christine Frum, James G. Pfaus, and James W. Lewis, "The Common Neural Bases Between Sexual Desire and Love: A Multilevel Kernel Density fMRI Analysis," *Journal of Sexual Medicine* 9, no. 4 (2012): 1048–54.

14. Elizabeth E. Bruch and M. E. J. Newman, "Aspirational Pursuit of Mates in Online Dating Markets," *Science Advances* 4, no. 8 (2018): eaap9815.

15. Eli J. Finkel, Paul W. Eastwick, Benjamin R. Karney, Harry T. Reis, and Susan Sprecher, "Online Dating: A Critical Analysis from the Perspective of Psychological Science," *Psychological Science in the Public Interest* 13, no. 1 (2012): 3–66; John T. Cacioppo, Stephanie Cacioppo, Gian C. Gonzaga, Elizabeth L. Ogburn, and Tyler J. VanderWeele, "Marital Satisfaction and Break-ups Differ Across On-Line and Off-Line Meeting Venues," *Proceedings of the National Academy of Sciences* 110, no. 25 (2013): 10135–40; Catalina L. Toma, "Online Dating and Psychological Wellbeing: A Social Compensation Perspective," *Current Opinion in Psychology* 46 (2022): 101331.

16. Michelle Drouin, *Out of Touch: How to Survive an Intimacy Famine* (Cambridge, MA: MIT Press, 2022); David J. Linden, *Touch: The Science of the Hand, Heart, and Mind* (New York: Penguin Books, 2016); Robin I. M. Dunbar, "The Social Role of Touch in Humans and Primates: Behavioural Function and Neurobiological Mechanisms," *Neuroscience and Biobehavioral Reviews* 34, no. 2 (2010): 260–68; Nina G. Jablonski, "Social and Affective Touch in Primates and Its Role in the Evolution of Social Cohesion," *Neuroscience* 464 (2021): 117–25; Marina Von Mohr, L. P. Kirsch, and Aikaterini

Fotopoulou, "The Soothing Function of Touch: Affective Touch Reduces Feelings of Social Exclusion," *Scientific Reports* 7, no. 1 (2017): 13516.
17. Hans R. IJzerman, *Heartwarming: How Our Inner Thermostat Made Us Human* (New York: W. W. Norton, 2021); Dunbar, "The Social Role of Touch"; Jablonski, "Social and Affective Touch in Primates."
18. Tiffany M. Field, ed., *Touch in Early Development* (New York: Psychology Press, 2014); Linden, *Touch*.
19. Haidt, *The Anxious Generation*.
20. Robert Bozick, "Is There Really a Sex Recession? Period and Cohort Effects on Sexual Inactivity Among American Men, 2006–2019," *American Journal of Men's Health* 15, no. 6 (2021): 15579883211057710; Jean M. Twenge, "Possible Reasons US Adults Are Not Having Sex as Much as They Used To," *JAMA Network Open* 3, no. 6 (2020): e203889; Jean M. Twenge, Ryne A. Sherman, and Brooke E. Wells, "Declines in Sexual Frequency Among American Adults, 1989–2014," *Archives of Sexual Behavior* 46, no. 8 (2017): 2389–401; Kate Julian, "Why Are Young People Having So Little Sex?," *The Atlantic*, December 2018.
21. Peter Ueda, Catherine H. Mercer, Christina Ghaznavi, and Debby Herbenick, "Trends in Frequency of Sexual Activity and Number of Sexual Partners Among Adults Aged 18 to 44 Years in the US, 2000–2018," *JAMA Network Open* 3, no. 6 (2020): e203833.

Chapter Two: Crave

1. Justin R. Garcia and Helen E. Fisher, "Why We Hook Up: Searching for Sex or Looking for Love?," in *Gender, Sex, and Politics: In the Streets and Between the Sheets in the 21st Century*, ed. Shira Tarrant (New York: Routledge, 2015), 238–50; Helen E. Fisher and Justin R. Garcia, "Slow Love: Courtship in the Digital Age," in *The New Psychology of Love*, 2nd ed., ed. Robert J. Sternberg and Karin Sternberg (Cambridge: Cambridge University Press, 2018), 208–22.
2. Joshua M. Ackerman, Vladas Griskevicius, and Norman P. Li, "Let's Get Serious: Communicating Commitment in Romantic Relationships," *Journal of Personality and Social Psychology* 100, no. 6 (2011): 1079.
3. David M. Buss, "Human Mate Guarding," *Neuroendocrinology Letters* 23, no. 4 (2002): 23–29; Todd K. Shackelford, Aaron T. Goetz, F. E. Guta, and David P. Schmitt, "Mate Guarding and Frequent In-Pair Copulation in Humans: Concurrent or Compensatory Anti-Cuckoldry Tactics?," *Human Nature* 17 (2006): 239–52; Hanna Kokko and Liam J. Morrell, "Mate Guarding, Male Attractiveness, and Paternity Under Social Monogamy," *Behavioral Ecology* 16, no. 4 (2005): 724–31; Peter N. Brotherton and Petr E. Komers, "Mate Guarding and the Evolution of Social Monogamy in Mammals," in *Monogamy: Mating Strategies and Partnerships in Birds, Humans and Other Mammals*, ed. Ulrich H. Reichard and Christophe Boesch (Cambridge: Cambridge University Press, 2003), 42–58; Elaine Hatfield, Richard L. Rapson, and Luciana D. Martel, "Passionate Love and Sexual Desire," in *Handbook of Cultural Psychology*, ed. Shinobu Kitayama and Dov Cohen (New York: Guilford Press, 2007), 760–79.
4. Laura E. LeFebvre and Heather J. Carmack, "Catching Feelings: Exploring Commitment (Un)Readiness in Emerging Adulthood," *Journal of Social and Personal*

Relationships 37, no. 1 (2020): 143–62; William Jankowiak, "Mate Selection, Intimacy, and Marital Love in Chinese Society," in *Handbook on the Family and Marriage in China*, ed. Xiaowei Zang and Lucy X. Zhao (Cheltenham, UK: Edward Elgar Publishing, 2017), 53–74; Clément Feybesse and Elaine Hatfield, "Passionate Love," in *The New Psychology of Love*, 2nd ed., ed. Robert J. Sternberg and Karin Sternberg (Cambridge: Cambridge University Press, 2019), 183–207; Garcia and Fisher, "Why We Hook Up."

5. Bianca P. Acevedo, "The Positive Psychology of Romantic Love," in *Toward a Positive Psychology of Relationships: New Directions in Theory and Research*, ed. Meg A. Warren and Stewart I. Donaldson (Santa Barbara, CA: Praeger/ABC-CLIO, 2018), 55–75; Elaine Hatfield, Leslie Bensman, and Richard L. Rapson, "A Brief History of Social Scientists' Attempts to Measure Passionate Love," *Journal of Social and Personal Relationships* 29, no. 2 (2012): 143–64; Elaine Hatfield and Richard L. Rapson, "Historical and Cross-Cultural Perspectives on Passionate Love and Sexual Desire," *Annual Review of Sex Research* 4, no. 1 (1993): 67–97.

6. Helen E. Fisher, *Why We Love: The Nature and Chemistry of Romantic Love* (New York: Macmillan, 2004); Helen E. Fisher, Arthur Aron, and Lucy L. Brown, "Romantic Love: A Mammalian Brain System for Mate Choice," *Philosophical Transactions of the Royal Society B: Biological Sciences* 361, no. 1476 (2006): 2173–86.

7. William R. Jankowiak, ed., *Intimacies: Love and Sex Across Cultures* (New York: Columbia University Press, 2008); William R. Jankowiak and Edward F. Fischer, "A Cross-Cultural Perspective on Romantic Love," *Ethnology* 31, no. 2 (1992): 149–55; William Jankowiak and Alexandra J. Nelson, "The State of Ethnological Research on Love: A Critical Review," in *International Handbook of Love: Transcultural and Transdisciplinary Perspectives*, ed. Claude-Hélène Mayer and Elisabeth Vanderheiden (Cham, Switzerland: Springer, 2021), 23–39; William Jankowiak, "An Anthropologist Goes Looking for Love in All the Old Places: A Personal Account," in *The New Psychology of Love*, ed. Robert J. Sternberg and Karin Sternberg (Cambridge: Cambridge University Press, 2018), 240–58; William R. Jankowiak, "What Is This Thing Called Love?," *Emotion Review* 8, no. 2 (2016): 109–10; William Jankowiak, "Is the Pair Bond a Human Universal? An Analytical Essay," *International Review of Psychiatry* 35, no. 1 (2023): 16–24; William Jankowiak, *Illicit Monogamy: Inside a Fundamentalist Mormon Community* (New York: Columbia University Press, 2023); William Jankowiak, Yuzhou Shen, Shuhua Yao, Chang Wang, and Sarah Volsche, "Investigating Love's Universal Attributes: A Research Report from China," *Cross-Cultural Research* 49, no. 4 (2015): 422–36.

8. Jessica J. Hille, "Beyond Sex: A Review of Recent Literature on Asexuality," *Current Opinion in Psychology* 49 (2023): 101516; Jessica J. Hille, Lucy Bhuyan, and Heather Tillewein, "Ace and Poly: The Motivations and Experiences of People on the Ace Spectrum in Polyamorous Relationships," *Sexes* 5, no. 2 (2024): 111–19; Angela Chen, *Ace: What Asexuality Reveals About Desire, Society, and the Meaning of Sex* (Boston: Beacon Press, 2020).

9. C. Sue Carter and Lowell L. Getz, "Monogamy and the Prairie Vole," *Scientific American* 268, no. 6 (1993): 100–6; Lowell L. Getz and C. Sue Carter, "Prairie-Vole

Partnerships," *American Scientist* 84, no. 1 (1996): 56–62; C. Sue Carter and Lowell L. Getz, "Social and Hormonal Determinants of Reproductive Patterns in the Prairie Vole," in *Neurobiology: Current Comparative Approaches,* ed. David W. Pfaff (Berlin: Springer, 1985), 18–36.

10. C. Sue Carter, "Neuroendocrine Perspectives on Social Attachment and Love," *Psychoneuroendocrinology* 23, no. 8 (1998): 779–818; C. Sue Carter and Stephen W. Porges, "The Biochemistry of Love: An Oxytocin Hypothesis," *EMBO Reports* 14 (2013): 12–16; C. Sue Carter, "The Oxytocin and Vasopressin Pathway in the Context of Love and Fear," *Frontiers in Endocrinology* 8 (2017): 356; C. Sue Carter and Allison M. Perkeybile, "The Monogamy Paradox: What Do Love and Sex Have to Do with It?," *Frontiers in Ecology and Evolution* 6 (2018): 202; C. Sue Carter, "Sex, Love and Oxytocin: Two Metaphors and a Molecule," *Neuroscience and Biobehavioral Reviews* 143 (2022): 104948.

11. C. Sue Carter, William M. Kenkel, Emily L. MacLean, Stephen R. Wilson, Allison M. Perkeybile, Justin R. Yee, and Michael A. Kingsbury, "Is Oxytocin 'Nature's Medicine'?," *Pharmacological Reviews* 72, no. 4 (2020): 829–61; C. Sue Carter, "Oxytocin Pathways and the Evolution of Human Behavior," *Annual Review of Psychology* 65, no. 1 (2014): 17–39; Thomas R. Insel, "The Challenge of Translation in Social Neuroscience: A Review of Oxytocin, Vasopressin, and Affiliative Behavior," *Neuron* 65, no. 6 (2010): 768–79; Thomas R. Insel, "Toward a Neurobiology of Attachment," *Review of General Psychology* 4, no. 2 (2000): 176–85; Robert C. Froemke and Larry J. Young, "Oxytocin, Neural Plasticity, and Social Behavior," *Annual Review of Neuroscience* 44, no. 1 (2021): 359–81; Zachary R. Donaldson and Larry J. Young, "Oxytocin, Vasopressin, and the Neurogenetics of Sociality," *Science* 322, no. 5903 (2008): 900–4; Heather E. Ross and Larry J. Young, "Oxytocin and the Neural Mechanisms Regulating Social Cognition and Affiliative Behavior," *Frontiers in Neuroendocrinology* 30, no. 4 (2009): 534–47; Greg J. Norman et al., "Social Neuroscience: The Social Brain, Oxytocin, and Health," *Social Neuroscience* 7, no. 1 (2012): 18–29.

12. Joan B. Silk, "The Adaptive Value of Sociality in Mammalian Groups," *Philosophical Transactions of the Royal Society B: Biological Sciences* 362, no. 1480 (2007): 539–59; John C. Mitani, Josep Call, Peter M. Kappeler, Richard A. Palombit, and Joan B. Silk, eds., *The Evolution of Primate Societies* (Chicago: University of Chicago Press, 2012); Joseph Henrich, *The Secret of Our Success: How Culture Is Driving Human Evolution, Domesticating Our Species, and Making Us Smarter* (Princeton, NJ: Princeton University Press, 2016); David Brooks, *The Social Animal: The Hidden Sources of Love, Character, and Achievement* (New York: Random House, 2012); William von Hippel, *The Social Leap: The New Evolutionary Science of Who We Are, Where We Come From, and What Makes Us Happy* (New York: Harper, 2018).

13. Roy J. Levin and Wilma van Berlo, "Sexual Arousal and Orgasm in Subjects Who Experience Forced or Non-Consensual Sexual Stimulation: A Review," *Journal of Clinical Forensic Medicine* 11, no. 2 (2004): 82–88; Kelly D. Suschinsky and Meredith L. Chivers, "The Relationship Between Sexual Concordance and Orgasm Consistency in Women," *Journal of Sex Research* 55, no. 6 (2018): 704–18; Stanley Schachter and Jerome E. Singer, "Cognitive, Social, and Physiological Determinants of Emotional

State," *Psychological Review* 69 (1962): 379–99; Glenn L. White, Stanley Fishbein, and John Rutsein, "Passionate Love and the Misattribution of Arousal," *Journal of Personality and Social Psychology* 41, no. 1 (1981): 56–62.

14. Suschinsky and Chivers, "The Relationship Between Sexual Concordance and Orgasm Consistency in Women"; Schachter and Singer, "Cognitive, Social, and Physiological Determinants of Emotional State" White et al., "Passionate Love and the Misattribution of Arousal."
15. Carter et al., "Is Oxytocin 'Nature's Medicine'?"
16. Elaine Hatfield and Richard L. Rapson, *Love and Sex: Cross-Cultural Perspectives* (Boston: Allyn & Bacon, 1996); Peter B. Gray and Justin R. Garcia, *Evolution and Human Sexual Behavior* (Cambridge, MA: Harvard University Press, 2013); Helen E. Fisher, *Anatomy of Love: A Natural History of Mating, Marriage, and Why We Stray*, rev. ed. (New York: W. W. Norton, 2016); Stephanie Coontz, *Marriage, a History: How Love Conquered Marriage* (New York: Penguin, 2006).
17. Gray and Garcia, *Evolution and Human Sexual Behavior;* Fisher, *Anatomy of Love;* Robert Boyd, Joan B. Silk, and Kevin E. Langergraber, *How Humans Evolved,* 10th ed. (New York: W. W. Norton, 2023).
18. Sarah Blaffer Hrdy, *Mother Nature: A History of Mothers, Infants, and Natural Selection* (New York: Pantheon, 1999); Sarah Blaffer Hrdy, *The Woman That Never Evolved,* rev. ed. (Cambridge, MA: Harvard University Press, 1999); Sarah Blaffer Hrdy, *Mothers and Others: The Evolutionary Origins of Mutual Understanding* (Cambridge, MA: Belknap Press, 2011).
19. Helen E. Fisher, "Evolution of Human Serial Pairbonding," *American Journal of Physical Anthropology* 78, no. 3 (1989): 331–54; Helen E. Fisher, "The Nature and Evolution of Romantic Love," in *Romantic Passion: The Universal Experience?,* ed. William Jankowiak (New York: Columbia University Press, 1995), 23–41; Fisher, *Anatomy of Love;* Gray and Garcia, *Evolution and Human Sexual Behavior.*
20. Carter and Perkeybile, "The Monogamy Paradox"; Gray and Garcia, *Evolution and Human Sexual Behavior;* Catherine Salmon and Justin Hehman, "Social Versus Sexual Monogamy," in *The Oxford Handbook of Infidelity* (New York: Oxford University Press, 2022), 121–39; Dieter Lukas and Tim H. Clutton-Brock, "The Evolution of Social Monogamy in Mammals," *Science* 341, no. 6145 (2013): 526–30; Patricia Adair Gowaty, "Battles of the Sexes and Origins of Monogamy," *Oxford Ornithology Series* 6, no. 1 (1996): 21–52.
21. Christopher Ryan and Cacilda Jethá, *Sex at Dawn: How We Mate, Why We Stray, and What It Means for Modern Relationships* (New York: HarperCollins, 2012). Although I don't agree with their entire thesis, I support Ryan and Jethá's argument for the importance of the agricultural revolution to marriage and sex.
22. Colin R. Johnson, *Just Queer Folks: Gender and Sexuality in Rural America* (Philadelphia: Temple University Press, 2013).
23. Fisher, *Anatomy of Love;* Gray and Garcia, *Evolution and Human Sexual Behavior.*
24. Robert S. Walker, Kim R. Hill, Mark V. Flinn, and Ryan M. Ellsworth, "Evolutionary History of Hunter-Gatherer Marriage Practices," *PLOS One* 6, no. 4 (2011): e19066.
25. Elaine Hatfield, Yuching Mo, and Richard L. Rapson, "Love, Sex, and Marriage Across Cultures," in *The Oxford Handbook of Human Development and Culture: An*

Interdisciplinary Perspective, ed. Lene Arnett Jensen (New York: Oxford University Press, 2015), 570–85.
26. Hatfield and Rapson, *Love and Sex.*
27. Coontz, *Marriage, a History.*
28. Karl Grammer, Bernhard Fink, and Nick Neave, "Human Pheromones and Sexual Attraction," *European Journal of Obstetrics and Gynecology and Reproductive Biology* 118, no. 2 (2005): 135–42; Tristram D. Wyatt, "The Search for Human Pheromones: The Lost Decades and the Necessity of Returning to First Principles," *Proceedings of the Royal Society B: Biological Sciences* 282, no. 1804 (2015): 2014299; Randy Thornhill and Steven W. Gangestad, "The Scent of Symmetry: A Human Sex Pheromone That Signals Fitness?," *Evolution and Human Behavior* 20, no. 3 (1999): 175–201; Chen Oren, Liron Peled-Avron, and Simone G. Shamay-Tsoory, "A Scent of Romance: Human Putative Pheromone Affects Men's Sexual Cognition," *Social Cognitive and Affective Neuroscience* 14, no. 7 (2019): 719–26; Randy Thornhill et al., "Major Histocompatibility Complex Genes, Symmetry, and Body Scent Attractiveness in Men and Women," *Behavioral Ecology* 14, no. 5 (2003): 668–78; Tristram D. Wyatt, "Fifty Years of Pheromones," *Nature* 457, no. 7227 (2009): 262–63.
29. Susan M. Hughes, Mary A. Harrison, and Gordon G. Gallup Jr., "Sex Differences in Romantic Kissing Among College Students: An Evolutionary Perspective," *Evolutionary Psychology* 5, no. 3 (2007): 612–31.
30. William R. Jankowiak, Sarah L. Volsche, and Justin R. Garcia, "Is the Romantic-Sexual Kiss a Near Human Universal?," *American Anthropologist* 117, no. 3 (2015): 535–39; Frans B. M. de Waal, "The First Kiss: Foundations of Conflict Resolution Research in Animals," in *Natural Conflict Resolution,* ed. Filippo Aureli and Frans B. M. de Waal (Berkeley: University of California Press, 2000), 15–33; Colin A. Hendrie and Gayle Brewer, "Kissing as an Evolutionary Adaption to Protect Against Human Cytomegalovirus-Like Teratogenesis," *Medical Hypotheses* 74, no. 2 (2010): 222–24; Rafael Wlodarski and Robin I. M. Dunbar, "What's in a Kiss? The Effect of Romantic Kissing on Mate Desirability," *Evolutionary Psychology* 12, no. 1 (2014): 178–99.
31. John T. Cacioppo et al., "Marital Satisfaction and Break-ups Differ Across On-Line and Off-Line Meeting Venues," *Proceedings of the National Academy of Sciences* 110, no. 25 (2013): 10135–40; Jia Ming Hu, Rui Zhu, and Yiming Zhang, "Does Online Dating Make Relationships More Successful? Replication and Extension of a Previous Study," *Cyberpsychology, Behavior, and Social Networking* 27, no. 9 (2024): 635–40; Helen E. Fisher and Justin R. Garcia, "Mate Choice in the Digital Age," in *The Handbook of Human Mating,* ed. David M. Buss (New York: Oxford University Press, 2023), 777–95.
32. Eli Coleman, Eusebio Corona-Vargas, and Jessamyn V. Ford, "Advancing Sexual Pleasure as a Fundamental Human Right and Essential for Sexual Health, Overall Health and Well-Being: An Introduction to the Special Issue on Sexual Pleasure," *International Journal of Sexual Health* 33, no. 4 (2021): 473–77; Jessamyn V. Ford et al., "The World Association for Sexual Health's Declaration on Sexual Pleasure: A Technical Guide," *International Journal of Sexual Health* 33, no. 4 (2021): 612–42.
33. Justin R. Garcia, Elisabeth A. Lloyd, Kim Wallen, and Helen E. Fisher, "Variation in Orgasm Occurrence by Sexual Orientation in a Sample of US Singles," *Journal of Sexual*

Medicine 11, no. 11 (2014): 2645–52; Amanda N. Gesselman, Morgan Bennett-Brown, Sarah Dubé, Emily M. Kaufman, James T. Campbell, and Justin R. Garcia, "The Lifelong Orgasm Gap: Exploring Age's Impact on Orgasm Rates," *Sexual Medicine* 12, no. 3 (2024): qfae042; David A. Frederick, H. Kate St. John, Justin R. Garcia, and Elisabeth A. Lloyd, "Differences in Orgasm Frequency Among Gay, Lesbian, Bisexual, and Heterosexual Men and Women in a US National Sample," *Archives of Sexual Behavior* 47, no. 1 (2018): 273–88; Elisabeth A. Lloyd. *The Case of the Female Orgasm: Bias in the Science of Evolution* (Cambridge, MA: Harvard University Press, 2005); David A. Puts, Khytam Dawood, and Lisa L. Welling, "Why Women Have Orgasms: An Evolutionary Analysis," *Archives of Sexual Behavior* 41 (2012): 1127–43.

34. Gray and Garcia, *Evolution and Human Sexual Behavior;* Hatfield and Rapson, *Love and Sex;* Jankowiak, *Intimacies.*
35. Bonnie L. Hewlett and Barry S. Hewlett, "A Biocultural Approach to Sex, Love, and Intimacy in Central African Foragers and Farmers," in *Intimacies: Love and Sex Across Cultures,* ed. William R. Jankowiak (New York: Columbia University Press, 2008), 37–64; Barry S. Hewlett and Bonnie L. Hewlett, "Sex and Searching for Children Among Aka Foragers and Ngandu Farmers of Central Africa," *African Study Monographs* 31, no. 3 (2010): 107–25.
36. Cindy M. Meston and David M. Buss, "Why Humans Have Sex," *Archives of Sexual Behavior* 36 (2007): 477–507.

Chapter Three: Search

1. Paul Groves, "Leafy Sea Dragons," *Scientific American* 279, no. 6 (1998): 84–89; Kristy L. Forsgren and Christopher G. Lowe, "The Life History of Weedy Seadragons, *Phyllopteryx taeniolatus* (Teleostei: Syngnathidae)," *Marine and Freshwater Research* 57, no. 3 (2006): 313–22; Catherine Wallis, *Seahorses: Mysteries of the Oceans* (Charlestown, NH: Bunker Hill, 2004).
2. Margaret Bastock, *Courtship: An Ethological Study* (London: Routledge, 2018); Stephen M. Shuster and Michael John Wade, *Mating Systems and Strategies* (Princeton, NJ: Princeton University Press, 2003); Alan F. Dixson, *Mammalian Sexuality: The Act of Mating and the Evolution of Reproduction* (Cambridge: Cambridge University Press, 2021); Nathan H. Lents, *The Sexual Evolution: How 500 Million Years of Sex, Gender, and Mating Shape Modern Relationships* (New York: HarperCollins, 2025).
3. C. Tang Poy and Matthew H. Woolhouse, "The Attraction of Synchrony: A Hip-Hop Dance Study," *Frontiers in Psychology* 11 (2020): 588935; Julia F. Christensen, Camilo José Cela-Conde, and Antoni Gomila, "Not All About Sex: Neural and Biobehavioral Functions of Human Dance," *Annals of the New York Academy of Sciences* 1400, no. 1 (2017): 8–32; Karl Grammer, Kirsten B. Kruck, and Magnus S. Magnusson, "The Courtship Dance: Patterns of Nonverbal Synchronization in Opposite-Sex Encounters," *Journal of Nonverbal Behavior* 22, no. 1 (1998): 3–29; Bernhard Fink and Todd K. Shackelford, "Why Did Dance Evolve? A Comment on Laland, Wilkins, and Clayton (2016)," *Evolutionary Psychological Science* 3 (2017): 147–48; Kevin Laland, Clive Wilkins, and Nicky Clayton, "The Evolution of Dance," *Current Biology* 26 (2016): R5–R9.

4. Charlotte Duranton and Florence Gaunet, "Behavioural Synchronization from an Ethological Perspective: Overview of Its Adaptive Value," *Adaptive Behavior* 24, no. 3 (2016): 181–91; Oded Mayo and Ilanit Gordon, "In and Out of Synchrony—Behavioral and Physiological Dynamics of Dyadic Interpersonal Coordination," *Psychophysiology* 57, no. 6 (2020): e13574; Iain D. Couzin, "Synchronization: The Key to Effective Communication in Animal Collectives," *Trends in Cognitive Sciences* 22, no. 10 (2018): 844–46; Andrea Ravignani, Daniel L. Bowling, and W. Tecumseh Fitch, "Chorusing, Synchrony, and the Evolutionary Functions of Rhythm," *Frontiers in Psychology* 5 (2014): 1118; Jennifer L. Rotondo and Steven M. Boker, "Behavioral Synchronization in Human Conversational Interaction," in *Mirror Neurons and the Evolution of Brain and Language,* ed. Maksim Stamenov and Vittorio Gallese (Amsterdam: John Benjamins, 2008), 153–64; Daniel M. Fessler and Colin Holbrook, "Synchronized Behavior Increases Assessments of the Formidability and Cohesion of Coalitions," *Evolution and Human Behavior* 37, no. 6 (2016): 502–9; Leonie Koban, Anand Ramamoorthy, and Ivana Konvalinka, "Why Do We Fall into Sync with Others? Interpersonal Synchronization and the Brain's Optimization Principle," *Social Neuroscience* 14, no. 1 (2019): 1–9.
5. Mayo and Gordon, "In and Out of Synchrony"; Joana Coutinho, Alfredo Pereira, Patricia Oliveira-Silva, Deborah Meier, Vladimiro Lourenço, and Wolfgang Tschacher, "When Our Hearts Beat Together: Cardiac Synchrony as an Entry Point to Understand Dyadic Co-Regulation in Couples," *Psychophysiology* 58, no. 3 (2021): e13739; Emilio Ferrer and Jonathan L. Helm, "Dynamical Systems Modeling of Physiological Coregulation in Dyadic Interactions," *International Journal of Psychophysiology* 88, no. 3 (2013): 296–308; Stephen W. Porges, "The Vagal Paradox: A Polyvagal Solution," *Comprehensive Psychoneuroendocrinology* 16 (2023): 100200; Stephen W. Porges, "Polyvagal Theory: A Science of Safety," *Frontiers in Integrative Neuroscience* 16 (2022): 871227.
6. M. Cohen, M. Abargil, M. Ahissar, and S. Atzil, "Social and Nonsocial Synchrony Are Interrelated and Romantically Attractive," *Communications Psychology* 2, no. 1 (2024): 57.
7. William L. Yarber, Barbara W. Sayad, and Ryan Milhausen, *Human Sexuality: Diversity in Contemporary America* (New York: McGraw Hill, 2025).
8. James G. Pfaus, "Pathways of Sexual Desire," *Journal of Sexual Medicine* 6, no. 6 (2009): 1506–33; Lisa M. Diamond, "Emerging Perspectives on Distinctions Between Romantic Love and Sexual Desire," *Current Directions in Psychological Science* 13, no. 3 (2004): 116–19; Kristen P. Mark and Jenna A. Lasslo, "Maintaining Sexual Desire in Long-Term Relationships: A Systematic Review and Conceptual Model," *Journal of Sex Research* 55, nos. 4–5 (2018): 563–81.
9. Justin R. Garcia and Helen E. Fisher, "Why We Hook Up: Searching for Sex or Looking for Love?," in *Gender, Sex, and Politics: In the Streets and Between the Sheets in the 21st Century,* ed. Shira Tarrant (New York: Routledge, 2015), 238–50; Sarah G. Massey, Suzanne M. Seibold-Simpson, Amanda M. Merriwether, and Justin R. Garcia, "Hookup Culture, Sexual Victimization, and Sexual Health on American College Campuses," *Perspectives,* October 2015, 6–8; Justin R. Garcia, Chris Reiber, Sarah G. Massey, and Amanda M. Merriwether, "Sexual Hookup Culture: A Review," *Review of General Psychology* 16, no. 2 (2012): 161–76; Chris Reiber and Justin R. Garcia,

"Hooking Up: Gender Differences, Evolution, and Pluralistic Ignorance," *Evolutionary Psychology* 8, no. 3 (2010): 390–404; Justin R. Garcia and Chris Reiber, "Hook-Up Behavior: A Biopsychosocial Perspective," *Journal of Social, Evolutionary, and Cultural Psychology* 2, no. 4 (2008): 192–208; Justin R. Garcia, Amanda N. Gesselman, Sarah G. Massey, Suzanne M. Seibold-Simpson, and Amanda M. Merriwether, "Intimacy Through Casual Sex: Relational Context of Sexual Activity and Affectionate Behaviors," *Journal of Relationships Research* 9 (2018): e12, 1–10; Justin R. Garcia, Suzanne M. Seibold-Simpson, Sarah G. Massey, and Amanda M. Merriwether, "Casual Sex: Integrating Social, Behavioral, and Sexual Health Research," in *Handbook of the Sociology of Sexualities,* ed. John DeLamater and Rebecca F. Plante (Dordrecht: Springer, 2015), 203–22; Maryanne L. Fisher, Kayla Worth, Justin R. Garcia, and Tyler Meredith, "Feelings of Regret Following Uncommitted Sexual Encounters in Canadian University Students," *Culture, Health and Sexuality* 14, no. 1 (2012): 45–57; Helen E. Fisher and Justin R. Garcia, "Mate Choice in the Digital Age," in *The Oxford Handbook of Human Mating,* ed. David M. Buss (New York: Oxford University Press, 2023), 777–95.
10. Garcia et al., "Sexual Hookup Culture."
11. Garcia et al., "Casual Sex."
12. Garcia and Reiber, "Hook-up Behavior."
13. Garcia et al., "Sexual Hookup Culture."
14. Lisa Wade, *American Hookup: The New Culture of Sex on Campus* (New York: W. W. Norton, 2017).
15. Anne-Françoise Rutkowski and Carol Saunders, *Emotional and Cognitive Overload: The Dark Side of Information Technology* (London: Routledge, 2018); Glenn Geher and Scott Barry Kaufman, *Mating Intelligence Unleashed: The Role of the Mind in Sex, Dating, and Love* (Oxford: Oxford University Press, 2013); Tinke M. Pronk and Jaap J. A. Denissen, "A Rejection Mind-Set: Choice Overload in Online Dating," *Social Psychological and Personality Science* 11, no. 3 (2020): 388–96; Alison P. Lenton and Marco Francesconi, "How Humans Cognitively Manage an Abundance of Mate Options," *Psychological Science* 21, no. 4 (2010): 528–33.
16. David M. Buss, ed., *The Oxford Handbook of Human Mating* (New York: Oxford University Press, 2023).
17. Carl T. Bergstrom and Lee Alan Dugatkin, *Evolution,* 3rd ed. (New York: W. W. Norton, 2023).
18. Norman P. Li, James M. Bailey, Douglas T. Kenrick, and Joanna A. W. Linsenmeier, "The Necessities and Luxuries of Mate Preferences: Testing the Tradeoffs," *Journal of Personality and Social Psychology* 82, no. 6 (2002): 947–55; Melissa R. Fales, David A. Frederick, Justin R. Garcia, Kelly A. Gildersleeve, Martie G. Haselton, and Helen E. Fisher, "Mating Markets and Bargaining Hands: Mate Preferences for Attractiveness and Resources in Two National US Studies," *Personality and Individual Differences* 88 (2016): 78–87; Elizabeth E. Bruch and M. E. J. Newman, "Structure of Online Dating Markets in US Cities," *Sociological Science* 6 (2019): 219–34; Roy F. Baumeister, Tiffany Reynolds, Bo Winegard, and Kathleen D. Vohs, "Competing for Love: Applying Sexual Economics Theory to Mating Contests," *Journal of Economic Psychology* 63 (2017):

230–41; Kathleen D. Vohs and Roy F. Baumeister, "Correcting Some Misrepresentations About Gender and Sexual Economics Theory: Comment on Rudman and Fetterolf (2014)," *Psychological Science* 26, no. 9 (2015): 1522–23.
19. David M. Buss and David P. Schmitt, "Mate Preferences and Their Behavioral Manifestations," *Annual Review of Psychology* 70, no. 1 (2019): 77–110; Gert Stulp, Abraham P. Buunk, Thomas V. Pollet, Daniel Nettle, and Sara Verhulst, "Are Human Mating Preferences with Respect to Height Reflected in Actual Pairings?," *PLOS One* 8, no. 1 (2013): e54186; Alexandre Courtiol, Michel Raymond, Bernard Godelle, and Jean-Baptiste Ferdy, "Mate Choice and Human Stature: Homogamy as a Unified Framework for Understanding Mating Preferences," *Evolution* 64 (2010): 2189–203; Rebecca Sear, "Height and Reproductive Success: How a Gambian Population Compares with the West," *Human Nature* 17 (2006): 405–18; Daniel Nettle, "Women's Height, Reproductive Success and the Evolution of Sexual Dimorphism in Modern Humans," *Proceedings of the Royal Society B: Biological Sciences* 269, no. 1503 (2002): 1919–23; Gert Stulp, Sara Verhulst, Thomas V. Pollet, and Abraham P. Buunk, "The Effect of Female Height on Reproductive Success Is Negative in Western Populations, but More Variable in Non-Western Populations," *American Journal of Human Biology* 24, no. 4 (2012): 486–94.
20. Bruch and Newman, "Aspirational Pursuit of Mates in Online Dating Markets."
21. David Conroy-Beam, Cari D. Goetz, and David M. Buss, "What Predicts Romantic Relationship Satisfaction and Mate Retention Intensity: Mate Preference Fulfillment or Mate Value Discrepancies?," *Evolution and Human Behavior* 37, no. 6 (2016): 440–48.
22. Baumeister et al., "Competing for Love"; Roy F. Baumeister and Kathleen D. Vohs, "Sexual Economics: Sex as Female Resource for Social Exchange in Heterosexual Interactions," *Personality and Social Psychology Review* 8, no. 4 (2004): 339–63; Roy F. Baumeister and Kathleen D. Vohs, "Sexual Economics, Culture, Men, and Modern Sexual Trends," *Society* 49 (2012): 520–24; Roy F. Baumeister and Jean M. Twenge, "Cultural Suppression of Female Sexuality," *Review of General Psychology* 6 (2002): 166–203.
23. Lena Klümper, Mathias Hassebrauck, and Sascha Schwarz, "Intersexual and Intrasexual Differences in Mate Selection Preferences Among Lesbian Women, Gay Men, and Bisexual Women and Men," *Archives of Sexual Behavior* 53, no. 1 (2024): 177–203; Richard A. Lippa, "The Preferred Traits of Mates in a Cross-National Study of Heterosexual and Homosexual Men and Women: An Examination of Biological and Cultural Influences," *Archives of Sexual Behavior* 36 (2007): 193–208; Kyle L. Gobrogge, Paul S. Perkins, Jessica H. Baker, Kristen D. Balcer, S. Marc Breedlove, and Kelly L. Klump, "Homosexual Mating Preferences from an Evolutionary Perspective: Sexual Selection Theory Revisited," *Archives of Sexual Behavior* 36 (2007): 717–23; Maria Scheller, Ana A. de Sousa, Lori A. Brotto, and Anthony C. Little, "The Role of Sexual and Romantic Attraction in Human Mate Preferences," *Journal of Sex Research* 61, no. 2 (2024): 299–312; Amber J. Coventry, Sarah Mixner, Benjamin Gelbart, K. V. Walter, Daniel Conroy-Beam, and Thomas C. German, "Deconfounding Sex and Sex of Partner in Mate-Preference Research," *Psychological Science* 36, no. 2 (2025): 116–29.

24. Coventry et al., "Deconfounding Sex and Sex of Partner in Mate-Preference Research."
25. J. Michael Bailey, Patrick Y. Kim, Adam Hills, and James A. W. Linsenmeier, "Butch, Femme, or Straight Acting? Partner Preferences of Gay Men and Lesbians," *Journal of Personality and Social Psychology* 73, no. 5 (1997): 960–73.
26. Amanda Shea and Jen Bell, "We Asked 64,000 Women What They Look for in a Partner. The Most Important Thing? Kindness," *Clue*, July 24, 2019, https://helloclue.com/articles/sex/idealpartner; Laura J. Botzet, Amanda Shea, Virginia J. Vitzthum, Anna Druet, Maddie Sheesley, and Tanja M. Gerlach, "The Link Between Age and Partner Preferences in a Large, International Sample of Single Women," *Human Nature* 34, no. 5 (2023): 539–68.
27. Lippa, "The Preferred Traits of Mates in a Cross-National Study of Heterosexual and Homosexual Men and Women."
28. Melissa R. Fales, David A. Frederick, Justin R. Garcia, Kelly A. Gildersleeve, Martie G. Haselton, and Helen E. Fisher, "Mating Markets and Bargaining Hands: Mate Preferences for Attractiveness and Resources in Two National US Studies," *Personality and Individual Differences* 88 (2016): 78–87.
29. Emily K. Hughes, Erica B. Slotter, and Lisa F. Emery, "Expanding Me, Loving Us: Self-Expansion Preferences, Experiences, and Romantic Relationship Commitment," *Self and Identity* 22, no. 2 (2023): 227–46; Arthur Aron, Gary Lewandowski, Benjamin Branand, Debra Mashek, and Elaine Aron, "Self-Expansion Motivation and Inclusion of Others in Self: An Updated Review," *Journal of Social and Personal Relationships* 39, no. 12 (2022): 3821–52; Renee N. Balzarini, Aditi Sharma, and Amy Muise, "Virtually Connected: Do Shared Novel Activities in Virtual Reality Enhance Self-Expansion and Relationship Quality?," *Behavioral Sciences* 15, no. 1 (2025): 67.
30. Geoffrey Miller, *The Mating Mind: How Sexual Choice Shaped the Evolution of Human Nature* (New York: Anchor, 2001).
31. Peter M. Todd, Lars Penke, Barbara Fasolo, and Alison P. Lenton, "Different Cognitive Processes Underlie Human Mate Choices and Mate Preferences," *Proceedings of the National Academy of Sciences of the United States of America* 104, no. 38 (2007): 15011–16.
32. Robert L. Trivers, "Parental Investment and Sexual Selection," in *Sexual Selection and the Descent of Man, 1871–1971,* ed. Bernard Campbell (Chicago: Aldine, 1972), 136–79; Michael J. Wade and Stephen M. Shuster, "The Evolution of Parental Care in the Context of Sexual Selection: A Critical Reassessment of Parental Investment Theory," *American Naturalist* 160, no. 3 (2002): 285–92; Hillard Kaplan, "A Theory of Fertility and Parental Investment in Traditional and Modern Human Societies," *American Journal of Physical Anthropology* 101, suppl. 23 (1996): 91–135; Elisabeth Oberzaucher, "Parental Investment Theory," in *Encyclopedia of Evolutionary Psychological Science,* ed. Todd K. Shackelford and Viviana A. Weekes-Shackelford (Cham, Switzerland: Springer, 2021).
33. Bram Neyt, Stijn Vandenbulcke, and Stijn Baert, "Are Men Intimidated by Highly Educated Women? Undercover on Tinder," *Economics of Education Review* 73 (2019): 101914.
34. David Conroy-Beam and David M. Buss, "How Are Mate Preferences Linked with Actual Mate Selection? Tests of Mate Preference Integration Algorithms Using Computer Simulations and Actual Mating Couples," *PLOS One* 11, no. 6 (2016): e0156078.

35. Peter B. Gray and Justin R. Garcia, *Evolution and Human Sexual Behavior* (Cambridge, MA: Harvard University Press, 2013).
36. Eli J. Finkel et al., "The Suffocation Model: Why Marriage in America Is Becoming an All-or-Nothing Institution," *Current Directions in Psychological Science* 24, no. 3 (2015): 238–44; Eli J. Finkel, *The All-or-Nothing Marriage: How the Best Marriages Work* (New York: Penguin, 2019).
37. Peter K. Jonason, Justin R. Garcia, Helen E. Fisher, et al., "Relationship Dealbreakers: Traits People Avoid in Potential Mates," *Personality and Social Psychology Bulletin* 41, no. 12 (2015): 1697–711.
38. Amanda N. Gesselman, Gregory D. Webster, and Justin R. Garcia, "Has Virginity Lost Its Virtue? Relationship Stigma Associated with Being a Sexually Inexperienced Adult," *Journal of Sex Research* 54, no. 2 (2017): 202–13.

Chapter Four: Date

1. Donald G. Dutton and Arthur P. Aron, "Some Evidence for Heightened Sexual Attraction Under Conditions of High Anxiety," *Journal of Personality and Social Psychology* 30, no. 4 (1974): 510–17; Helen E. Fisher, *Why We Love: The Nature and Chemistry of Romantic Love* (New York: Macmillan, 2004).
2. Moira Weigel, *Labor of Love: The Invention of Dating* (New York: Farrar, Straus and Giroux, 2017).
3. Beth L. Bailey, *From Front Porch to Back Seat: Courtship in Twentieth-Century America* (Baltimore: Johns Hopkins University Press, 1989).
4. Dutton and Aron, "Some Evidence for Heightened Sexual Attraction Under Conditions of High Anxiety."
5. Stephen W. Porges, *The Polyvagal Theory: Neurophysiological Foundations of Emotions, Attachment, Communication, and Self-Regulation* (New York: W. W. Norton, 2011).
6. Stephen W. Porges, "Neuroception: A Subconscious System for Detecting Threats and Safety," *Zero to Three* 24, no. 5 (2004): 19–24; Porges, *The Polyvagal Theory*.
7. Mandy Len Catron, "To Fall in Love with Anyone, Do This," *New York Times*, January 9, 2015.
8. Arthur Aron et al., "The Experimental Generation of Interpersonal Closeness: A Procedure and Some Preliminary Findings," *Personality and Social Psychology Bulletin* 23, no. 4 (1997): 363–77.
9. John Gottman, Julie Schwartz Gottman, Doug Abrams, and Rachel Carlton Abrams, *Eight Dates: Essential Conversations for a Lifetime of Love* (New York: Workman, 2019).
10. Kelsey Campbell, Lissette C. Silva, and David W. Wright, "Rituals in Unmarried Couple Relationships: An Exploratory Study," *Family and Consumer Sciences Research Journal* 40, no. 1 (2011): 45–57.
11. Esther Perel, *Mating in Captivity* (New York: HarperCollins, 2006); John M. Gottman, "Gottman Method Couple Therapy," *Clinical Handbook of Couple Therapy* 4, no. 8 (2008): 138–64; John Gottman, Julie Gottman, and Mary A. McNulty, "The Role of Trust and Commitment in Love Relationships," in *Foundations for Couples' Therapy,* ed. Jennifer Fitzgerald (New York: Routledge, 2017), 438–52.

12. Jessica E. French, Leah J. Bolton, and Andrea L. Meltzer, "Virtual Speed Dating: Utilizing Online-Meeting Platforms to Study Initial Attraction and Relationship Formation," *Personal Relationships* 31, no. 2 (2024): 420–44.
13. Helen E. Fisher and Justin R. Garcia, "Mate Choice in the Digital Age," in *The Handbook of Human Mating,* ed. David M. Buss (New York: Oxford University Press, 2023), 777–95.
14. Carolina Thomé, "Hierarchies in Heterosexuality: Orgasms, Intercourse and Sexual Scripts," *Sexualities* 28, nos. 1–2 (2023): 283–300; Casey E. Copen, Anjani Chandra, and Gladys Martinez, "Prevalence and Timing of Oral Sex with Opposite-Sex Partners Among Females and Males Aged 15–24 Years: United States, 2007–2010," National Health Statistics Reports, no. 56, U.S. Department of Health and Human Services, August 16, 2012; Caroline Heldman and Lisa Wade, "Hook-Up Culture: Setting a New Research Agenda," *Sexuality Research and Social Policy* 7 (2010): 323–33.
15. Helen E. Fisher and Justin R. Garcia, "Slow Love: Courtship in the Digital Age," in *The New Psychology of Love,* 2nd ed., ed. Robert J. Sternberg and Karin Sternberg (Cambridge: Cambridge University Press, 2018), 208–22; Helen E. Fisher, "Slow Love: How Casual Sex May Be Improving Marriage," *Nautilus,* March 5, 2015, https://nautil.us/slow-love-how-casual-sex-may-be-improving-marriage-236359/.
16. Jean M. Twenge, Ryne A. Sherman, and Brooke E. Wells, "Declines in Sexual Frequency Among American Adults, 1989–2014," *Archives of Sexual Behavior* 46, no. 8 (2017): 2389–401; Jean M. Twenge, "Possible Reasons US Adults Are Not Having Sex as Much as They Used To," *JAMA Network Open* 3, no. 6 (2020): e203889; Kate Julian, "The Sex Recession," *The Atlantic,* December 2018.
17. Justin R. Garcia and Helen E. Fisher, "Why We Hook Up: Searching for Sex or Looking for Love?," in *Gender, Sex, and Politics: In the Streets and Between the Sheets in the 21st Century,* ed. Shira Tarrant (New York: Routledge, 2015), 238–50.
18. William G. Eberhard, "Copulatory Courtship and Cryptic Female Choice in Insects," *Biological Reviews* 66, no. 1 (1991): 1–31; Geoffrey Miller, *The Mating Mind: How Sexual Choice Shaped the Evolution of Human Nature* (New York: Anchor, 2001).
19. Fisher and Garcia, "Slow Love."
20. Andreas Wilke, John M. Hutchinson, Peter M. Todd, and Daniel J. Kruger, "Is Risk Taking Used as a Cue in Mate Choice?" *Evolutionary Psychology* 4, no. 1 (2006); Karolina Sylwester and Bogusław Pawłowski, "Daring to Be Darling: Attractiveness of Risk Takers as Partners in Long- and Short-Term Sexual Relationships," *Sex Roles* 64 (2011): 695–706; Virginia R. Henderson et al., "When Risky Is Attractive: Sensation Seeking and Romantic Partner Selection," *Personality and Individual Differences* 38, no. 2 (2005): 311–25.
21. Alexandra Eleftheriou et al., "Does Attractiveness Influence Condom Use Intentions in Heterosexual Men? An Experimental Study," *BMJ Open* 6, no. 6 (2016): e010883; Alexandra Eleftheriou et al., "Does Attractiveness Influence Condom Use Intentions in Women Who Have Sex with Men?," *PLOS One* 14, no. 5 (2019): e0217152; Terri D. Conley et al., "Condom Use Errors Among Sexually Unfaithful and Consensually Nonmonogamous Individuals," *Sexual Health* 10, no. 5 (2013): 463–64.
22. Fisher and Garcia, "Slow Love"; Fisher and Garcia, "Mate Choice in the Digital Age."

23. Justin R. Garcia et al., "Sexting Among Singles in the U.S.A.: Prevalence of Sending, Receiving, and Sharing Sexual Messages and Images," *Sexual Health* 13, no. 5 (2016): 428–35.
24. Amanda S. Marcotte, Amanda N. Gesselman, Helen E. Fisher, and Justin R. Garcia, "Women's and Men's Reactions to Receiving Unsolicited Genital Images from Men," *Journal of Sex Research* 58, no. 4 (2021): 512–21.
25. Garcia et al., "Sexting Among Singles in the U.S.A."; Michelle Drouin, Megan Coupe, and Jeff R. Temple, "Is Sexting Good for Your Relationship? It Depends...," *Computers in Human Behavior* 75 (2017): 749–56.
26. Charlene Olckers and Marné Hattingh, "The Dark Side of Social Media—Cyberbullying, Catfishing and Trolling: A Systematic Literature Review," *EPiC Series in Computing* 84 (2022): 71–86; Ryan D. Lamphere and Katherine T. Lucas, "Online Romance in the 21st Century: Deceptive Online Dating, Catfishing, Romance Scams, and 'Mail Order' Marriages," in *Handbook of Research on Deception, Fake News, and Misinformation Online*, ed. Melissa Zimdars and Kembrew McLeod (Hershey, PA: IGI Global, 2019), 475–88.
27. Steven Petrow, "My Girlfriend and I Meow at Each Other. It's Not as Unusual as It Sounds," *Washington Post,* August 28, 2017; Nikki Thorburn, "What's the Deal with Baby Talk in Relationships?," *Fashion Journal,* August 16, 2022, https://fashionjournal.com.au/life/baby-talk-in-relationships/.
28. Chrisanna Northrup, Pepper Schwartz, and James Witte, *The Normal Bar: The Surprising Secrets of Happy Couples and What They Reveal About Creating a New Normal in Your Relationship* (New York: Harmony, 2013).
29. Lauren M. Papp, Jennifer Danielewicz, and Crystal Cayemberg, "Are We Facebook Official? Implications of Dating Partners' Facebook Use and Profiles for Intimate Relationship Satisfaction," *Cyberpsychology, Behavior, and Social Networking* 15, no. 2 (2012): 85–90; Brittany L. Lane, Charles W. Piercy, and Caleb T. Carr, "Making It Facebook Official: The Warranting Value of Online Relationship Status Disclosures on Relational Characteristics," *Computers in Human Behavior* 56 (2016): 1–8; Brady Robards and Sian Lincoln, "Making It 'Facebook Official': Reflecting on Romantic Relationships Through Sustained Facebook Use," *Social Media + Society* 2, no. 4 (2016).
30. Liesel L. Sharabi and Allison Hopkins, "Picture Perfect? Examining Associations Between Relationship Quality, Attention to Alternatives, and Couples' Activities on Instagram," *Journal of Social and Personal Relationships* 38, no. 12 (2021): 3518–542.

Chapter Five: Mate

1. Ron Charles, "The Kinsey Institute Has Lost State Funding. Hoosiers Should Be Embarrassed," *Washington Post,* May 10, 2023; Suzette Hackney, "Kinsey Institute and Indiana University Split Would Damage Both," *Indianapolis Star,* March 7, 2024.
2. Charles R. Berger, "Goals, Plans, and Mutual Understanding in Relationships," in *Individuals in Relationships,* ed. Steve Duck (Thousand Oaks, CA: SAGE, 1993), 30–59; Franziska Denzinger, Sabine Backes, and Veronika Brandstätter, "Same Same but Different: Similarity of Goals and Implicit Motives in Intimate Relationships," *Motivation Science* 4, no. 1 (2018): 60–77; Kristin Laurin et al., "Power and the Pursuit of a Partner's Goals," *Journal of Personality and Social Psychology* 110, no. 6 (2016): 840–68.

3. Arthur Aron and Elaine N. Aron, "Self and Self-Expansion in Relationships," in *Knowledge Structures in Close Relationships,* ed. Garth J. O. Fletcher and Julie Fitness (New York: Psychology Press, 2014), 325–44; Arthur Aron, Elaine N. Aron, and Danny Smollan, "Self-Expansion Motivation and Inclusion of Others in Self: An Updated Review," *Journal of Social and Personal Relationships* 39, no. 12 (2022): 3821–52; Elaine N. Aron and Arthur Aron, "Love and Expansion of the Self: The State of the Model," *Personal Relationships* 3, no. 1 (1996): 45–58; Arthur Aron and Elaine N. Aron, *Love and the Expansion of Self: Understanding Attraction and Satisfaction* (New York: Hemisphere, 1986); Arthur Aron and Elaine N. Aron, "Romantic Relationships from the Perspectives of the Self-Expansion Model and Attachment Theory: Partially Overlapping Circles," in *Dynamics of Romantic Love: Attachment, Caregiving, and Sex,* ed. Mario Mikulincer and Gail S. Goodman (New York: Guilford Press, 2006), 359–82.
4. Arthur Aron, Elaine N. Aron, and Danny Smollan, "Inclusion of Other in the Self Scale and the Structure of Interpersonal Closeness," *Journal of Personality and Social Psychology* 63, no. 4 (1992): 596–612; Christopher R. Agnew et al., "Thinking Close: Measuring Relational Closeness as Perceived Self–Other Inclusion," in *Handbook of Closeness and Intimacy,* ed. Debra Mashek and Arthur Aron (New York: Psychology Press, 2004), 113–26.
5. Brian Parkinson, Greg Simons, and Karen Niven, "Sharing Concerns: Interpersonal Worry Regulation in Romantic Couples," *Emotion* 16, no. 4 (2016): 449–58.
6. Helen E. Fisher and Justin R. Garcia, "Slow Love: Courtship in the Digital Age," in *The New Psychology of Love,* 2nd ed., ed. Robert J. Sternberg and Karin Sternberg (Cambridge: Cambridge University Press, 2018), 208–22.
7. Fisher and Garcia, "Slow Love."
8. Peter B. Gray and Justin R. Garcia, *Evolution and Human Sexual Behavior* (Cambridge, MA: Harvard University Press, 2013).
9. Gray and Garcia, *Evolution and Human Sexual Behavior;* Roy F. Baumeister, Tiffany Reynolds, Bo Winegard, and Kathleen D. Vohs, "Competing for Love: Applying Sexual Economics Theory to Mating Contests," *Journal of Economic Psychology* 63 (2017): 230–41.
10. Melvin Konner, *The Evolution of Childhood: Relationships, Emotion, Mind* (Cambridge, MA: Harvard University Press, 2011).
11. Saloni Dattani, Lucas Rodés-Guirao, and Max Roser, "Fertility Rate," Our World in Data, 2025, https://ourworldindata.org/fertility-rate.
12. R. Rugh, *The Frog: Its Reproduction and Development* (Charleston, SC: Legare Street Press, 2022).
13. C. Sarabian, V. Curtis, and R. McMullan, "Evolution of Pathogen and Parasite Avoidance Behaviours," *Philosophical Transactions of the Royal Society B: Biological Sciences* 373, no. 1751 (2018): 20170256; J. M. Ackerman, "Disease Avoidance Hypothesis," in *Encyclopedia of Evolutionary Psychological Science,* ed. Todd K. Shackelford and Viviana A. Weekes-Shackelford (Cham, Switzerland: Springer, 2021), 2044–50.
14. Mark Schaller and L. A. Duncan, "The Behavioral Immune System: Its Evolution and Social Psychological Implications," in *Evolution and the Social Mind*, 293–307 (New York: Psychology Press, 2011); Joshua M. Ackerman, Sarah E. Hill, and Damian R.

Murray, "The Behavioral Immune System: Current Concerns and Future Directions," *Social and Personality Psychology Compass* 12, no. 2 (2018): e12371.
15. Randolph M. Nesse, "Natural Selection and the Regulation of Defenses: A Signal Detection Analysis of the Smoke Detector Principle," *Evolution and Human Behavior* 26 (2005): 88–105; Stephen C. Stearns and Ruslan Medzhitov, *Evolutionary Medicine* (Oxford: Oxford University Press, 2024).
16. Jennifer S. Barber, William Miller, Yasamin Kusunoki, Sarah R. Hayford, and Kelly B. Guzzo, "Intimate Relationship Dynamics and Changing Desire for Pregnancy Among Young Women," *Perspectives on Sexual and Reproductive Health* 51, no. 3 (2019): 143–52.
17. Gray and Garcia, *Evolution and Human Sexual Behavior;* Helen E. Fisher, *Anatomy of Love: A Natural History of Mating, Marriage, and Why We Stray,* rev. ed. (New York: W. W. Norton, 2016).
18. Gray and Garcia, *Evolution and Human Sexual Behavior.*
19. İsmail Guvensoy and Gaye Erdem, "The Effects of Ideal Standards and Parental Approval on Mate Choice Among Emerging Adults," *Journal of Social and Personal Relationships* 40, no. 1 (2023): 174–200; Jeffry H. Larson, Marietta Malnar, and Dean Busby, "Factors Related to Perceived Parental Approval of Adult Sons' and Daughters' Heterosexual Engagements," *Journal of Family Therapy* 38, no. 3 (2016): 364–85; Preshous Benjamin, Vincent Ngafeeson, and Jared A. Durtschi, "Can Family Approval in Early Marriage Predict Early Divorce?," *Journal of Couple and Relationship Therapy,* published online April 4, 2025.
20. Gregory L. Schlomer, Marco Del Giudice, and Bruce J. Ellis, "Parent–Offspring Conflict Theory: An Evolutionary Framework for Understanding Conflict Within Human Families," *Psychological Review* 118, no. 3 (2011): 496–521; Catherine A. Salmon and James Malcolm, "Parent-Offspring Conflict," in *The Oxford Handbook of Evolutionary Family Psychology,* ed. Catherine A. Salmon and Todd K. Shackelford (Oxford: Oxford University Press, 2011), 83–96.
21. Anthony A. Volk, "Human Breastfeeding Is Not Automatic: Why That's So and What It Means for Human Evolution," *Journal of Social, Evolutionary, and Cultural Psychology* 3, no. 4 (2009): 305–14; Takumi Tsutaya and Nozomi Mizushima, "Evolutionary Biological Perspectives on Current Social Issues of Breastfeeding and Weaning," *American Journal of Biological Anthropology* 181 (2023): 81–93.
22. Erika C. Odom, Ruowei Li, Kelley S. Scanlon, Cria G. Perrine, and Laurence Grummer-Strawn, "Reasons for Earlier than Desired Cessation of Breastfeeding," *Pediatrics* 131, no. 3 (2013): e726–e732.
23. Sarah Blaffer Hrdy, *Mothers and Others: The Evolutionary Origins of Mutual Understanding* (Cambridge, MA: Belknap Press of Harvard University Press, 2011).
24. Tamas Bereczkei, Petra Gyuris, and Glenn E. Weisfeld, "Sexual Imprinting in Human Mate Choice," *Proceedings of the Royal Society B: Biological Sciences* 271, no. 1544 (2004): 1129–34; T. Bereczkei, P. Gyuris, P. Koves, and L. Bernath, "Homogamy, Genetic Similarity, and Imprinting: Parental Influence on Mate Choice Preferences," *Personality and Individual Differences* 33, no. 5 (2002): 677–90; Saori Nojo, Satoshi Tamura, and Yasuo Ihara, "Human Homogamy in Facial Characteristics: Does a Sexual-Imprinting-Like Mechanism Play a Role?," *Human Nature* 23, no. 3 (2012): 323–340.

25. Bereczkei et al., "Homogamy, Genetic Similarity, and Imprinting."
26. Peter B. Gray, Shelly L. Volsche, Justin R. Garcia, and Helen E. Fisher, "The Roles of Pet Dogs and Cats in Human Courtship and Dating," *Anthrozoös* 28, no. 4 (2015): 673–83.
27. Abraham P. Buunk, Justin H. Park, and Lesley A. Duncan, "Cultural Variation in Parental Influence on Mate Choice," *Cross-Cultural Research* 44, no. 1 (2010): 23–40; Carin Perilloux, Diana S. Fleischman, and David M. Buss, "The Daughter-Guarding Hypothesis: Parental Influence on, and Emotional Reactions to, Offspring's Mating Behavior," *Evolutionary Psychology* 6, no. 2 (2008); M. V. Flinn, "Paternal Care in a Caribbean Village," in *Father-Child Relations: Cultural and Biosocial Contexts,* ed. Barry S. Hewlett (New York: Routledge, 2010); 57–84; David C. Geary and M. V. Flinn, "Evolution of Human Parental Behavior and the Human Family," *Parenting* 1, nos. 1–2 (2001): 5–61; Rebecca Sear, "Family and Fertility: Does Kin Help Influence Women's Fertility, and How Does This Vary Worldwide?," *Population Horizons* 14, no. 1 (2018): 18–34.
28. Bobbi S. Low, *Why Sex Matters: A Darwinian Look at Human Behavior,* rev. ed. (Princeton, NJ: Princeton University Press, 2015).
29. Alfred C. Kinsey, Wardell B. Pomeroy, and Clyde E. Martin, *Sexual Behavior in the Human Male* (Philadelphia: W. B. Saunders, 1948); Alfred C. Kinsey, Wardell B. Pomeroy, Clyde E. Martin, and Paul H. Gebhard, *Sexual Behavior in the Human Female* (Philadelphia: W. B. Saunders, 1953).
30. See the website of the Kinsey Institute, https://kinseyinstitute.org/.
31. Amanda N. Gesselman, Margaret Bennett-Brown, Simon Dubé, Ellen M. Kaufman, Jessica T. Campbell, and Justin R. Garcia, "The Lifelong Orgasm Gap: Exploring Age's Impact on Orgasm Rates," *Sexual Medicine* 12, no. 3 (2024): qfae042; Justin R. Garcia, Elisabeth A. Lloyd, Kim Wallen, and Helen E. Fisher, "Variation in Orgasm Occurrence by Sexual Orientation in a Sample of US Singles," *Journal of Sexual Medicine* 11, no. 11 (2014): 2645–52; Grace M. Wetzel, Diana T. Sanchez, Amanda N. Gesselman, Olivia R. Adams, James T. Campbell, and Justin R. Garcia, "Exploring the Orgasm Gap Across Racial/Ethnic Groups: A Descriptive Analysis," *Journal of Sex Research* 62, no. 5 (2025): 683–92; David A. Frederick, H. Kate St. John, Justin R. Garcia, and Elisabeth A. Lloyd, "Differences in Orgasm Frequency Among Gay, Lesbian, Bisexual, and Heterosexual Men and Women in a US National Sample," *Archives of Sexual Behavior* 47, no. 1 (2018): 273–88; Julia R. Heiman, "Orgasmic Disorders in Women," in *Principles and Practice of Sex Therapy*, 4th ed., ed. Sandra Risa Leiblum (New York: Guilford Press 2007), 84–123; Elisabeth A. Lloyd, *The Case of the Female Orgasm: Bias in the Science of Evolution* (Cambridge, MA: Harvard University Press, 2006); Gray and Garcia, *Evolution and Human Sexual Behavior*.
32. Barry R. Komisaruk, Carlos Beyer-Flores, and Beverly Whipple, *The Science of Orgasm* (Baltimore: Johns Hopkins University Press, 2006); Elisabeth A. Lloyd, *The Case of the Female Orgasm: Bias in the Science of Evolution* (Cambridge, MA: Harvard University Press, 2006); David A. Puts, Khaled Dawood, and Lisa L. Welling, "Why Women Have Orgasms: An Evolutionary Analysis," *Archives of Sexual Behavior* 41 (2012): 1127–43; Kim Wallen, P. Z. Myers, and Elisabeth A. Lloyd, "Zietsch and Santtila's Study Is Not Evidence Against the By-Product Theory of Female Orgasm," *Animal*

Behaviour 84, no. 5 (2012): e1; Susana Basanta and Laura Nuño de la Rosa, "The Female Orgasm and the Homology Concept in Evolutionary Biology," *Journal of Morphology* 284, no. 1 (2023): e21544; Vincent J. Lynch, "Clitoral and Penile Size Variability Are Not Significantly Different: Lack of Evidence for the By-Product Theory of the Female Orgasm," *Evolution and Development* 10, no. 4 (2008): 396–97; Brendan P. Zietsch and Pekka Santtila, "Genetic Analysis of Orgasmic Function in Twins and Siblings Does Not Support the By-Product Theory of Female Orgasm," *Animal Behaviour* 82, no. 5 (2011): 1097–101; Kim Wallen and Elisabeth A. Lloyd, "Female Sexual Arousal: Genital Anatomy and Orgasm in Intercourse," *Hormones and Behavior* 59, no. 5 (2011): 780–92; Elisabeth A. Lloyd, "Orgasms and Objectification," *Archives of Sexual Behavior* 46 (2017): 1191–94.

33. Alan F. Dixson, *Sexual Selection and the Origins of Human Mating Systems* (Oxford: Oxford University Press, 2009); Gray and Garcia, *Evolution and Human Sexual Behavior* (Cambridge, MA: Harvard University Press, 2013); Holly Dunsworth, "Sexual Evolution: Why We Have Sex and What It Has to Do with Human Origins," *Sapiens,* September 23, 2016, https://www.sapiens.org/biology/sexual-evolution-pleasure/.

34. Rhonda N. Balzarini and Amy Muise, "Beyond the Dyad: A Review of the Novel Insights Gained from Studying Consensual Non-Monogamy," *Current Sexual Health Reports* 12 (2020): 398–404; David L. Rodrigues, "A Narrative Review of the Dichotomy Between the Social Views of Non-Monogamy and the Experiences of Consensual Non-Monogamous People," *Archives of Sexual Behavior* 53, no. 3 (2024): 931–40; M. L. Haupert, Amy C. Moors, Amanda N. Gesselman, and Justin R. Garcia, "Estimates and Correlates of Engagement in Consensually Non-Monogamous Relationships," *Current Sexual Health Reports* 9 (2017): 155–65; Amy C. Moors, Michelle Denise Vaughan, Sharon M. Flicker, Kaiponanea T. Matsumura, and Heath A. Schechinger, "Advancing Consensual Non-Monogamy in Psychological Research, Practice, and Policy: A Guide for Psychologists," *Social Issues and Policy Review* 19, no. 1 (2025): e12108; Amy C. Moors, "Five Misconceptions About Consensually Nonmonogamous Relationships," *Current Directions in Psychological Science* 32, no. 5 (2023): 355–61; Amy C. Moors, Dylan F. Selterman, and Terri D. Conley, "Personality Correlates of Desire to Engage in Consensual Non-Monogamy Among Lesbian, Gay, and Bisexual Individuals," *Journal of Bisexuality* 17, no. 4 (2017): 418–34; M. L. Haupert, Amanda N. Gesselman, Amy C. Moors, Helen E. Fisher, and Justin R. Garcia, "Prevalence of Experiences with Consensual Nonmonogamous Relationships: Findings from Two National Samples of Single Americans," *Journal of Sex and Marital Therapy* 43, no. 5 (2017): 424–40; Shelby Astle, Kara Langin, Jared R. Anderson, and Amy C. Moors, "Understanding Relationship Labels: A Content Analysis of Consensual Non-Monogamous Relationship Agreements," *Sexuality and Culture* 28, no. 2 (2024): 710–32; Joel Anderson, Alena Bondarchuk-McLaughlin, Scarlet Rosa, Karen D. Goldschlager, and D. X. Hinton Jordan, "A Qualitative Exploration of the Experiences of Disclosing Non-Monogamy," *Archives of Sexual Behavior* 54 (2025): 1481–95; Rhonda N. Balzarini, Lorne Campbell, Taylor Kohut, Bjarne M. Holmes, Justin J. Lehmiller, Jennifer J. Harman, and Nicole Atkins, "Perceptions of Primary and Secondary Relationships in Polyamory," *PLOS One* 12, no. 5 (2017): e0177841.

35. Christopher M. Gleason, *American Poly: A History* (New York: Oxford University Press, 2024); Deborah Anapol, *Polyamory in the 21st Century: Love and Intimacy with Multiple Partners* (Lanham, MD: Rowman & Littlefield, 2010).
36. James Fenske, "African Polygamy: Past and Present," *Journal of Development Economics* 117 (2015): 58–73; Tsoaledi Daniel Thobejane and Takayindisa Flora, "An Exploration of Polygamous Marriages: A Worldview," *Mediterranean Journal of Social Sciences* 5, no. 27 (2014): 1058–66; Sumanto Al Qurtuby, "Between Polygyny and Monogamy: Marriage in Saudi Arabia and Beyond," *Al-Jāmi'ah: Journal of Islamic Studies* 60, no. 1 (2022).
37. Aghaghia Rahimzadeh, "Fraternal Polyandry and Land Ownership in Kinnaur, Western Himalaya," *Human Ecology* 48, no. 5 (2020): 573–84; Melvyn C. Goldstein, "Fraternal Polyandry and Fertility in a High Himalayan Valley in Northwest Nepal," *Human Ecology* 4, no. 3 (1976): 223–33; Gerald D. Berreman, "Himalayan Polyandry and the Domestic Cycle," *American Ethnologist* 2, no. 1 (1975): 127–38; Robert S. Walker, Mark V. Flinn, and Kim R. Hill, "Evolutionary History of Partible Paternity in Lowland South America," *Proceedings of the National Academy of Sciences* 107, no. 45 (2010): 19195–200; Stephen Beckerman, Roberto Lizarralde, Carol Ballew, Sissel Schroeder, Christina Fingelton, Angela Garrison, and Helen Smith, "The Bari Partible Paternity Project: Preliminary Results," *Current Anthropology* 39, no. 1 (1998): 164–68; Stephen Beckerman, Manuel Lizarralde, Daniela Peluso, Cédric Yvinec, Nathan Harris, Daniel Parker, Robert Walker, and Kim Hill, "Partible Paternity, the Secondary Sex Ratio, and a Possible Trivers-Willard Effect," *Current Anthropology* 58, no. 4 (2017): 540–43; Stephen Beckerman and Paul Valentine, "Introduction: The Concept of Partible Paternity Among Native South Americans," in *Cultures of Multiple Fathers: The Theory and Practice of Partible Paternity in Lowland South America,* ed. Stephen Beckerman and Paul Valentine (Gainesville: University Press of Florida, 2002), 1–13.
38. Haupert et al., "Prevalence of Experiences with Consensual Nonmonogamous Relationships."
39. M. L. Haupert, Amy C. Moors, Amanda N. Gesselman, and Justin R. Garcia, "Estimates and Correlates of Engagement in Consensually Non-Monogamous Relationships."
40. Amy C. Moors, "Has the American Public's Interest in Information Related to Relationships Beyond 'The Couple' Increased over Time?," *Journal of Sex Research* 54, no. 6 (2017): 677–84.
41. Amy C. Moors, Heath A. Schechinger, Rhonda Balzarini, and Sharon Flicker, "Internalized Consensual Non-Monogamy Negativity and Relationship Quality Among People Engaged in Polyamory, Swinging, and Open Relationships," *Archives of Sexual Behavior* 50, no. 4 (2021): 1389–400; Moors, "Five Misconceptions About Consensually Nonmonogamous Relationships"; Alicia N. Rubel and Anthony F. Bogaert, "Consensual Nonmonogamy: Psychological Well-Being and Relationship Quality Correlates," *Journal of Sex Research* 52, no. 9 (2015): 961–82; Joel R. Anderson, Jordan D. X. Hinton, Alena Bondarchuk-McLaughlin, Scarlet Rosa, Kian Jin Tan, and Lily Moor, "Countering the Monogamy-Superiority Myth: A Meta-Analysis of the Differences in Relationship Satisfaction and Sexual Satisfaction as a Function of Relationship Orientation," *Journal of Sex Research,* published online March 24, 2025.

42. Justin J. Lehmiller, *Tell Me What You Want: The Science of Sexual Desire and How It Can Help You Improve Your Sex Life* (New York: Balance Press, 2018).
43. Benjamin Le, Nicole L. Dove, Christopher R. Agnew, Michelle S. Korn, and Anna A. Mutso, "Predicting Nonmarital Romantic Relationship Dissolution: A Meta-Analytic Synthesis," *Personal Relationships* 17, no. 3 (2010): 377–90.
44. Sandra Murray and John G. Holmes, "A Leap of Faith? Positive Illusions in Romantic Relationships," *Personality and Social Psychology Bulletin* 23 (1997): 586–604; Caryl E. Rusbult, Paul A. M. Van Lange, Tim Wildschut, Nicole A. Yovetich, and Julie Verette, "Perceived Superiority in Close Relationships: Why It Exists and Persists," *Journal of Personality and Social Psychology* 79 (2000): 521–45; Le et al., "Predicting Nonmarital Romantic Relationship Dissolution."
45. Aron et al., "Inclusion of Other in the Self Scale and the Structure of Interpersonal Closeness"; Agnew et al., "Thinking Close."
46. Bianca P. Acevedo, Arthur Aron, Helen E. Fisher, and Lucy L. Brown, "Neural Correlates of Long-Term Intense Romantic Love," *Social Cognitive and Affective Neuroscience* 7, no. 2 (2012): 145–59.
47. Jon K. Maner, Debra A. Rouby, and Gian C. Gonzaga, "Automatic Inattention to Attractive Alternatives: The Evolved Psychology of Relationship Maintenance," *Evolution and Human Behavior* 29, no. 5 (2008): 343–49.

Chapter Six: Nest

1. Brian J. Willoughby, Jason S. Carroll, and D. M. Busby, "The Different Effects of 'Living Together': Determining and Comparing Types of Cohabiting Couples," *Journal of Social and Personal Relationships* 29, no. 3 (2012): 397–419; Pamela J. Smock and Wendy D. Manning, "Cohabiting Partners' Economic Circumstances and Marriage," *Demography* 34 (1997): 331–41; Adrianne Evans and Edith Gray, "Cross-National Differences in Income Pooling Among Married and Cohabiting Couples," *Journal of Marriage and Family* 83, no. 2 (2021): 534–50; Mariagrazia Bosisio, Natalie O. Rosen, Josée Dubé, Marie-Pier Vaillancourt-Morel, Marie-Ève Daspe, and Sophie Bergeron, "Will You Be Happy for Me? Associations Between Self-Reported, Perceived, and Observed Responses to Positive Events and Sexual Well-Being in Cohabiting Couples," *Journal of Social and Personal Relationships* 39, no. 8 (2022): 2454–77; W. Gibb Dyer, Scott Kofford, and Brian J. Willoughby, "Cohabiting Couples: A Neglected Family Form That Is Important for the Field of Family Business," *European Journal of Family Business* 13, no. 2 (2023): 137–48; Wendy D. Manning, Krista K. Westrick-Payne, and Gary J. Gates, "Cohabitation and Marriage Among Same-Sex Couples in the 2019 ACS and CPS: A Research Note," *Demography* 59, no. 5 (2022): 1595–605; Randi Zahl-Olsen, Frode Thuen, and Tor Håkon Stea, "Cohabitation, Marriage, and Union Dissolution in Norway: A Comparative Prospective Study," *Journal of Divorce and Remarriage* 64, nos. 5–6 (2023): 199–216; Katharina Richter, Susanne Adam, Lisa Geiss, Laura Peter, and Georg Niklewski, "Two in a Bed: The Influence of Couple Sleeping and Chronotypes on Relationship and Sleep. An Overview," *Chronobiology International* 33, no. 10 (2016): 1464–72.
2. Barnard College Archives, "The LeClair Affair," n.d., accessed April 23, 2025, https://barnardarchives.omeka.net/exhibits/show/introduction/leclairaffair/page1; Shawn

Cohen, "50 Years Since the LeClair Affair: Is Living Together Outside of Marriage a Problem?," The Society Pages, March 27, 2018, https://thesocietypages.org/ccf/2018/03/27/50-years-since-the-leclair-affair-is-living-together-outside-of-marriage-a-problem/; Estelle B. Freedman, "Coming of Age at Barnard, 1968," *The Sixties* 1, no. 2 (2008): 209–22.

3. Barbara Perelli-Harris and Bin Kuang, "Demographic Perspectives on Cohabitation," in *Research Handbook on Marriage, Cohabitation and the Law,* ed. Rebecca Probert and Sharon Thompson (Cheltenham, UK: Edward Elgar Publishing, 2024), 296–312.

4. Yuan Mei Yang, "Premarital Cohabitation and the Gendered Division of Household Labor in China," *Chinese Journal of Sociology* 10, no. 2 (2024): 274–312; House of Commons Library, "Research Briefing SN03372: Cohabitation and the Law," last modified April 9, 2021, https://commonslibrary.parliament.uk/research-briefings/sn03372/; Julianna Menasce Horowitz, Nikki Graf, and Gretchen Livingston, "The Landscape of Marriage and Cohabitation in the U.S.," Pew Research Center, November 6, 2019, https://www.pewresearch.org/social-trends/2019/11/06/the-landscape-of-marriage-and-cohabitation-in-the-u-s/.

5. Horowitz et al., "The Landscape of Marriage and Cohabitation in the U.S."

6. Antony W. Dnes, "Marriage, Cohabitation, and Same-Sex Marriage," *Independent Review* 12, no. 1 (2007): 85–99; Brian Powell, Natasha Y. Quadlin, and Oren Pizmony-Levy, "Public Opinion, the Courts, and Same-Sex Marriage: Four Lessons Learned," *Social Currents* 2, no. 1 (2015): 3–12.

7. I. Anna S. Olsson and Karolina Westlund, "More than Numbers Matter: The Effect of Social Factors on Behaviour and Welfare of Laboratory Rodents and Non-Human Primates," *Applied Animal Behaviour Science* 103, nos. 3–4 (2007): 229–54.

8. Kim Hill, "Life History Theory and Evolutionary Anthropology," *Evolutionary Anthropology: Issues, News, and Reviews* 2, no. 3 (1993): 78–88; Marco Del Giudice, Hillard S. Kaplan, and Steven W. Gangestad, "Life History Theory and Evolutionary Psychology," in *The Handbook of Evolutionary Psychology,* ed. David M. Buss (Hoboken, NJ: John Wiley & Sons, 2015), 68–95; Stephen C. Stearns and Álvaro M. Rodrigues, "On the Use of 'Life History Theory' in Evolutionary Psychology," *Evolution and Human Behavior* 41, no. 6 (2020): 474–85; Daniel Nettle and Willem E. Frankenhuis, "Life-History Theory in Psychology and Evolutionary Biology: One Research Programme or Two?," *Philosophical Transactions of the Royal Society B: Biological Sciences* 375, no. 1803 (2020): 20190490; Steven C. Hertler, Aurelio J. Figueredo, Mateo Peñaherrera-Aguirre, Heitor B. Fernandes, and Michael A. Woodley, *Life History Evolution: A Biological Meta-Theory for the Social Sciences* (Cham, Switzerland: Springer, 2018).

9. Frank W. Marlowe, "Hunter-Gatherers and Human Evolution," *Evolutionary Anthropology* 14, no. 2 (2005): 54–67; Peter B. Gray and Justin R. Garcia, *Evolution and Human Sexual Behavior* (Cambridge, MA: Harvard University Press, 2013).

10. Rebecca Sear, David W. Lawson, Hillard Kaplan, and Mary K. Shenk, "Understanding Variation in Human Fertility: What Can We Learn from Evolutionary Demography?," *Philosophical Transactions of the Royal Society B: Biological Sciences* 371, no. 1692 (2016): 20150144.

11. World Population Review, "Total Fertility Rate by Country," accessed April 21, 2025, https://worldpopulationreview.com/country-rankings/total-fertility-rate.

12. Rebecca Sear, "The Impact of Reproduction on Gambian Women: Does Controlling for Phenotypic Quality Reveal Costs of Reproduction?," *American Journal of Physical Anthropology* 132, no. 4 (2007): 632–41; Sear et al., "Understanding Variation in Human Fertility"; Ruth Mace and Rebecca Sear, "Birth Interval and the Sex of Children in a Traditional African Population: An Evolutionary Analysis," *Journal of Biosocial Science* 29, no. 4 (1997): 499–507.
13. Helen E. Fisher, *The Sex Contract: The Evolution of Human Behavior* (New York: William Morrow, 1982).
14. Kimberly Coe, *The Ancestress Hypothesis: Visual Art as Adaptation* (New Brunswick, NJ: Rutgers University Press, 2003); Kimberly Coe, Nathan E. Aiken, and Charles T. Palmer, "Once Upon a Time: Ancestors and the Evolutionary Significance of Stories," *Anthropological Forum* 16, no. 1 (March 2006): 21–40.
15. Lawrence Kilham, "Cooperative Breeding of American Crows," *Journal of Field Ornithology* 55, no. 3 (1984): 349–56; Sarah Blaffer Hrdy, *Mothers and Others: The Evolutionary Origins of Mutual Understanding* (Cambridge, MA: Belknap Press, 2011).
16. Paul L. Vasey and David P. VanderLaan, "Fa'afafine," in *Encyclopedia of Evolutionary Psychological Science,* ed. Todd K. Shackelford and Viviana A. Weekes-Shackelford (Cham, Switzerland: Springer, 2021), 2875–76; Paul L. Vasey and Nicholas H. Bartlett, "What Can the Samoan 'Fa'afafine' Teach Us About the Western Concept of Gender Identity Disorder in Childhood?," *Perspectives in Biology and Medicine* 50, no. 4 (2007): 481–90.
17. Paul L. Vasey, David S. Pocock, and Doug P. VanderLaan, "Kin Selection and Male Androphilia in Samoan Fa'afafine," *Evolution and Human Behavior* 28, no. 3 (2007): 159–67; David P. VanderLaan, David L. Forrester, L. J. Petterson, and Paul L. Vasey, "Offspring Production Among the Extended Relatives of Samoan Men and Fa'afafine," *PLOS One* 7, no. 4 (2012): e36088.
18. Dritan Mazrekaj, Kristof De Witte, and Svenja Cabus, "School Outcomes of Children Raised by Same-Sex Parents: Evidence from Administrative Panel Data," *American Sociological Review* 85, no. 5 (2020): 830–56; Simona Cheng and Brian Powell, "Measurement, Methods, and Divergent Patterns: Reassessing the Effects of Same-Sex Parents," *Social Science Research* 52 (2015): 615–26; Angela Crowl, Shanondora Ahn, and Jeanine Baker, "A Meta-Analysis of Developmental Outcomes for Children of Same-Sex and Heterosexual Parents," *Journal of GLBT Family Studies* 4, no. 3 (2008): 385–407.
19. Maria Pallotta-Chiarolli, Penny Haydon, and Anna Hunter, "'These Are Our Children': Polyamorous Parenting," in *LGBT-Parent Families: Innovations in Research and Implications for Practice,* ed. Abbie E. Goldberg and Katherine R. Allen (New York: Springer, 2013), 117–31; Amy C. Moors, "Five Misconceptions About Consensually Nonmonogamous Relationships," *Current Directions in Psychological Science* 32, no. 5 (2023): 355–61.
20. Riana Minocher, Monique Borgerhoff Mulder, and Cody T. Ross, "Little Evidence That Nonmonogamous Family Structures Are Detrimental to Children's Well-Being in Mpimbwe, Tanzania," *Proceedings of the National Academy of Sciences* 121, no. 52 (2024): e2407785121.
21. Fisher, *The Sex Contract;* Rebecca Sear, "The Male Breadwinner Nuclear Family Is Not the 'Traditional' Human Family, and Promotion of This Myth May Have Adverse

Health Consequences," *Philosophical Transactions of the Royal Society B: Biological Sciences* 376, no. 1827 (2021): 20200020.
22. Shing Chan, Hsiu-Ling Liao, and Kevin O'Regan, "Moving Back: Spatial and Demographic Differences in Boomeranging to Parents," *Journal of Urban Affairs* (2025): 1–35.
23. LendingTree, "Survey: Young Adults Who Moved Back Home During Pandemic," accessed April 21, 2025, https://www.lendingtree.com/home/mortgage/living-with-parents-survey/.
24. LendingTree, "Survey."
25. Paul F. Hemez, Chanell N. Washington, and Rose M. Kreider, "America's Families and Living Arrangements: 2022," Current Population Reports P20-587, US Census Bureau, Washington, DC, 2024, https://www2.census.gov/library/publications/2024/demo/p20-587.pdf; "Parent Trap: Nearly Half of Adult Gen Zers Getting Financial Help from Mom and Dad, According to BofA Study," PR Newswire, July 10, 2024, https://www.prnewswire.com/news-releases/parent-trap-nearly-half-of-adult-gen-zers-getting-financial-help-from-mom-and-dad-according-to-bofa-study-302192841.html.
26. Martin V. Flinn, "Evolutionary Anthropology of the Human Family," in *The Oxford Handbook of Evolutionary Family Psychology,* ed. Catherine Salmon and Todd K. Shackelford (Oxford: Oxford University Press, 2011), 12–32; Catherine Salmon and Todd K. Shackelford, eds., *The Oxford Handbook of Evolutionary Family Psychology* (Oxford: Oxford University Press, 2011).
27. Elaine Hatfield, "Passionate Love, Companionate Love, and Intimacy," in *Intimacy,* ed. Martin Fisher and George Stricker (Boston: Springer, 1982), 267–92; Elaine C. Hatfield, Julie T. Pillemer, Megan U. O'Brien, and Yuliana C. L. Le, "The Endurance of Love: Passionate and Companionate Love in Newlywed and Long-Term Marriages," *Interpersona: An International Journal on Personal Relationships* 2, no. 1 (2008): 35–64; David A. Frederick, Janet Lever, Brian J. Gillespie, and Justin R. Garcia, "What Keeps Passion Alive? Sexual Satisfaction Is Associated with Sexual Communication, Mood Setting, Sexual Variety, Oral Sex, Orgasm, and Sex Frequency in a National US Study," *Journal of Sex Research* 54, no. 2 (2017): 186–201; Bianca P. Acevedo and Arthur Aron, "Does a Long-Term Relationship Kill Romantic Love?," *Review of General Psychology* 13, no. 1 (2009): 59–65.
28. Hatfield, "Passionate Love, Companionate Love, and Intimacy"; Hatfield et al., "The Endurance of Love"; Elaine Hatfield and G. William Walster, *A New Look at Love* (Lanham, MD: University Press of America, 1985).
29. Frederick et al., "What Keeps Passion Alive?"
30. David A. Frederick, Brooke J. Gillespie, Janet Lever, Vanessa Berardi, and Justin R. Garcia, "Debunking Lesbian Bed Death: Using Coarsened Exact Matching to Compare Sexual Practices and Satisfaction of Lesbian and Heterosexual Women," *Archives of Sexual Behavior* 50 (2021): 3601–19.
31. David Frederick, Brooke J. Gillespie, Janet Lever, Vanessa Berardi, and Justin R. Garcia, "Sexual Practices and Satisfaction Among Gay and Heterosexual Men in Romantic Relationships: A Comparison Using Coarsened Exact Matching in a US National Sample," *Journal of Sex Research* 58, no. 5 (2021): 545–59.
32. "Singles Pandemic: Why Global Birthrates Are Falling—and What It Means for Our Future," posted April 9, 2025, by Newsweek, YouTube, https://www.youtube.com

/watch?v=SXRelDqs2BE; United Nations, *World Fertility 2024*, UN DESA/POP /2024/TR/NO. 11 (New York: United Nations, 2025).

33. Sear, "The Male Breadwinner Nuclear Family Is Not the 'Traditional' Human Family"; Rebecca Sear, "Beyond the Nuclear Family: An Evolutionary Perspective on Parenting," *Current Opinion in Psychology* 7 (2016): 98–103.

34. Vivian P. Ta, Amanda N. Gesselman, Brea L. Perry, Helen E. Fisher, and Justin R. Garcia, "Stress of Singlehood: Marital Status, Domain-Specific Stress, and Anxiety in a National US Sample," *Journal of Social and Clinical Psychology* 36, no. 6 (2017): 461–85.

35. Amanda N. Gesselman, Carol Y. Franco, Elizabeth M. Brogdon, Peter B. Gray, Justin R. Garcia, and Helen E. Fisher, "Perceptions of Married Life Among Single Never-Married, Single Ever-Married, and Married Adults," *Personal Relationships* 26, no. 4 (2019): 586–601; Yuthika U. Girme, Yoojin Park, and Geoff MacDonald, "Coping or Thriving? Reviewing Intrapersonal, Interpersonal, and Societal Factors Associated with Well-Being in Singlehood from a Within-Group Perspective," *Perspectives on Psychological Science* 18, no. 5 (2023): 1097–120; Ta et al., "Stress of Singlehood."

36. Bella DePaulo, "Single and Flourishing: Transcending the Deficit Narratives of Single Life," *Journal of Family Theory and Review* 15, no. 3 (2023): 389–411.

37. Susan L. Brown and Alan Booth, "Cohabitation Versus Marriage: A Comparison of Relationship Quality," *Journal of Marriage and the Family* (1996): 668–78; Michael J. Rosenfeld and Katharina Roesler, "Cohabitation Experience and Cohabitation's Association with Marital Dissolution," *Journal of Marriage and Family* 81, no. 1 (2019): 42–58; Wendy D. Manning and Jessica A. Cohen, "Premarital Cohabitation and Marital Dissolution: An Examination of Recent Marriages," *Journal of Marriage and Family* 74, no. 2 (2012): 377–87; Larry L. Bumpass, James A. Sweet, and Andrew Cherlin, "The Role of Cohabitation in Declining Rates of Marriage," *Journal of Marriage and the Family* (1991): 913–27; Zhenchao Mu, "Premarital Cohabitation, Marital Dissolution, and Marital Quality in China," *Journal of Family Issues* 45, no. 5 (2024): 1305–25; Judith P. Soons and Matthijs Kalmijn, "Is Marriage More than Cohabitation? Well-Being Differences in 30 European Countries," *Journal of Marriage and Family* 71, no. 5 (2009): 1141–57; Sharon Sassler and Daniel T. Lichter, "Cohabitation and Marriage: Complexity and Diversity in Union-Formation Patterns," *Journal of Marriage and Family* 82, no. 1 (2020): 35–61.

38. Michael J. Rosenfeld and Katharina Roesler, "Cohabitation Experience and Cohabitation's Association with Marital Dissolution," *Journal of Marriage and Family* 81, no. 1 (2019): 42–58.

39. Sassler and Lichter, "Cohabitation and Marriage"; Andrew J. Cherlin, *The Marriage-Go-Round: The State of Marriage and the Family in America Today* (New York: Vintage, 2010).

40. Nathan D. Leonhardt, Natalie O. Rosen, Samantha J. Dawson, James J. Kim, Matthew D. Johnson, and Emily A. Impett, "Relationship Satisfaction and Commitment in the Transition to Parenthood: A Couple-Centered Approach," *Journal of Marriage and Family* 84, no. 1 (2022): 80–100.

41. Janet S. Hyde, John D. DeLamater, E. Ashby Plant, and Jennifer M. Byrd, "Sexuality During Pregnancy and the Year Postpartum," *Journal of Sex Research* 33, no. 2 (1996): 143–51; Holly L. McBride and J. Lynn Kwee, "Sex After Baby: Women's Sexual

Function in the Postpartum Period," *Current Sexual Health Reports* 9 (2017): 142–49; José R. Pauleta, Nuno M. Pereira, and Luís M. Graça, "Sexuality During Pregnancy," *Journal of Sexual Medicine* 7, no. 1 (2010): 136–42; Izabela Gałązka, Agnieszka Drosdzol-Cop, Beata Naworska, Mariola Czajkowska, and Violetta Skrzypulec-Plinta, "Changes in the Sexual Function During Pregnancy," *Journal of Sexual Medicine* 12, no. 2 (2015): 445–54; Güven Aslan, Dilek Aslan, Ayşe Kızılyar, Cenk Ispahi, and Ahmet Esen, "A Prospective Analysis of Sexual Functions During Pregnancy," *International Journal of Impotence Research* 17, no. 2 (2005): 154–57; E. Bartellas, J. M. Crane, M. Daley, K. A. Bennett, and D. Hutchens, "Sexuality and Sexual Activity in Pregnancy," *BJOG: An International Journal of Obstetrics and Gynaecology* 107, no. 8 (2000): 964–68; Mary Rowland, Laura Foxcroft, Wilma M. Hopman, and Rupa Patel, "Breastfeeding and Sexuality Immediately Postpartum," *Canadian Family Physician* 51, no. 10 (2005): 1366–67.

42. Sarah Blaffer Hrdy, *Mother Nature: A History of Mothers, Infants, and Natural Selection* (New York: Pantheon, 1999); Jennifer Hahn-Holbrook, Julianne Holt-Lunstad, Colin Holbrook, Sarah M. Coyne, and E. Thomas Lawson, "Maternal Defense: Breast Feeding Increases Aggression by Reducing Stress," *Psychological Science* 22, no. 10 (2011): 1288–95.

43. Deirdre O'Malley, Agnes Higgins, and Valerie Smith, "Exploring the Complexities of Postpartum Sexual Health," *Current Sexual Health Reports* 13 (2021): 128–35; Lawrence M. Leeman and Rebecca G. Rogers, "Sex After Childbirth: Postpartum Sexual Function," *Obstetrics and Gynecology* 119, no. 3 (2012): 647–55.

44. Peter B. Gray and Kermyt G. Anderson, *Fatherhood: Evolution and Human Paternal Behavior* (Cambridge, MA: Harvard University Press, 2010).

45. Peter B. Gray, Jason F. Chapman, T. Christopher Burnham, Matthew H. McIntyre, Susan F. Lipson, and Peter T. Ellison, "Human Male Pair Bonding and Testosterone," *Human Nature* 15 (2004): 119–31; Peter B. Gray, Sarah M. Kahlenberg, Elizabeth S. Barrett, Susan F. Lipson, and Peter T. Ellison, "Marriage and Fatherhood Are Associated with Lower Testosterone in Males," *Evolution and Human Behavior* 23, no. 3 (2002): 193–201; Peter B. Gray, Peter T. Ellison, and Benjamin C. Campbell, "Testosterone and Marriage Among Ariaal Men of Northern Kenya," *Current Anthropology* 48, no. 5 (2007): 750–55; Peter B. Gray, Jeremy Reece, Cheryl Coore-Desai, Tanesha Dinall, Simone Pellington, and Maureen Samms-Vaughan, "Testosterone and Jamaican Fathers: Exploring Links to Relationship Dynamics and Paternal Care," *Human Nature* 28 (2017): 201–18; Peter B. Gray, Chia-Feng Jeffrey Yang, and Harrison G. Pope Jr., "Fathers Have Lower Salivary Testosterone Levels than Unmarried Men and Married Non-Fathers in Beijing, China," *Proceedings of the Royal Society B: Biological Sciences* 273, no. 1584 (2006): 333–39.

46. Lee T. Gettler, Patty X. Kuo, Megha S. Sarma, Benjamin C. Trumble, Jennifer E. Burke Lefever, and Jean M. Braungart-Rieker, "Fathers' Oxytocin Responses to First Holding Their Newborns: Interactions with Testosterone Reactivity to Predict Later Parenting Behavior and Father-Infant Bonds," *Developmental Psychobiology* 63, no. 5 (2021): 1384–98.

47. Gettler et al., "Fathers' Oxytocin Responses to First Holding Their Newborns."

48. Peter B. Gray, Justin R. Garcia, Benjamin S. Crosier, and Helen E. Fisher, "Dating and Sexual Behavior Among Single Parents of Young Children in the United States," *Journal of Sex Research* 52, no. 2 (2015): 121–28.

49. Peter B. Gray, Chantal Franco, Justin R. Garcia, Amanda N. Gesselman, and Helen E. Fisher, "Romantic Dating Attitudes and Behaviors Among Single Parents in the U.S.," *Personal Relationships* 23, no. 3 (2016): 491–504; Gray et al., "Dating and Sexual Behavior Among Single Parents of Young Children in the United States."
50. Arlie Hochschild and Anne Machung, *The Second Shift: Working Families and the Revolution at Home* (New York: Penguin, 2012); Mary Blair-Loy, Arlie Hochschild, Annette J. Pugh, Joan C. Williams, and Heather Hartmann, "Stability and Transformation in Gender, Work, and Family: Insights from *The Second Shift* for the Next Quarter Century," *Community, Work and Family* 18, no. 4 (2015): 435–54.
51. Claire Cain Miller, "Nearly Half of Men Say They Do Most of the Home Schooling. 3 Percent of Women Agree," *New York Times*, May 6, 2020.
52. Miller, "Nearly Half of Men Say They Do Most of the Home Schooling."
53. W. Thomas Boyce, *The Orchid and the Dandelion: Why Sensitive Children Face Challenges and How All Can Thrive* (New York: Vintage, 2020).

Chapter Seven: Stray

1. William Jankowiak and Michael D. Hardgrave, "Individual and Societal Response to Sexual Betrayal: A View from Around the World," *Electronic Journal of Human Sexuality* 10 (2007): 1–7; Bram P. Buunk and Pieternel Dijkstra, "The Ultimate Betrayal? Infidelity and Solidarity in Close Relationships," in *Solidarity and Prosocial Behavior: An Integration of Sociological and Psychological Perspectives,* ed. Detlef Fetchenhauer, Andreas Flache, Abraham P. Buunk, and Siegwart Lindenberg (Boston: Springer, 2006), 111–24; Todd K. Shackelford, "Perceptions of Betrayal and the Design of the Mind," in *Evolutionary Social Psychology*, ed. Jeffry A. Simpson and Douglas T. Kenrick (New York: Psychology Press, 2013), 73–107; Susan L. Miller and Jon K. Maner, "Coping with Romantic Betrayal: Sex Differences in Responses to Partner Infidelity," *Evolutionary Psychology* 6, no. 3 (2008): 413–26; Beth Warach and Lawrence Josephs, "The Aftershocks of Infidelity: A Review of Infidelity-Based Attachment Trauma," *Sexual and Relationship Therapy* 36, no. 1 (2021): 68–90.
2. Don J. Sharpsteen and Lee A. Kirkpatrick, "Romantic Jealousy and Adult Romantic Attachment," *Journal of Personality and Social Psychology* 72, no. 3 (1997): 627–40; Ayala Malach Pines, *Romantic Jealousy: Causes, Symptoms, Cures* (New York: Routledge, 2016); Judith A. Easton, Lucas D. Schipper, and Todd K. Shackelford, "Morbid Jealousy from an Evolutionary Psychological Perspective," *Evolution and Human Behavior* 28, no. 6 (2007): 399–402; William Jankowiak, Monika Sudakov, and Benjamin C. Wilreker, "Co-Wife Conflict and Co-operation," *Ethnology* 44, no. 1 (2005): 81–98.
3. David M. Buss and Martie Haselton, "The Evolution of Jealousy," *Trends in Cognitive Sciences* 9, no. 11 (2005): 506; Christine R. Harris, "The Evolution of Jealousy: Did Men and Women, Facing Different Selective Pressures, Evolve Different 'Brands' of Jealousy? Recent Evidence Suggests Not," *American Scientist* 92, no. 1 (2004): 62–71; Jaak Panksepp, "The Evolutionary Sources of Jealousy: Cross-Species Approaches to Fundamental Issues," in *Handbook of Jealousy: Theory, Research, and Multidisciplinary Approaches,* ed. Sybil L. Hart and Maria Legerstee (Malden, MA: Wiley-Blackwell, 2010), 101–20; Jon K. Maner and Todd K. Shackelford, "The Basic Cognition of Jealousy: An Evolutionary Perspective," *European Journal of Personality* 22, no. 1 (2008): 31–36.

4. Devra G. Kleiman, "Monogamy in Mammals," *Quarterly Review of Biology* 52, no. 1 (1977): 39–69; Alan F. Dixson, *Primate Sexuality: Comparative Studies of the Prosimians, Monkeys, Apes, and Human Beings,* 2nd ed. (Oxford: Oxford University Press, 2012); Alan F. Dixson, *Mammalian Sexuality: The Act of Mating and the Evolution of Reproduction* (Cambridge: Cambridge University Press, 2021); Helen E. Fisher, *Anatomy of Love: A Natural History of Mating, Marriage, and Why We Stray* (New York: W. W. Norton, 1992); Peter B. Gray and Justin R. Garcia, *Evolution and Human Sexual Behavior* (Cambridge, MA: Harvard University Press, 2013); C. Sue Carter and Allison M. Perkeybile, "The Monogamy Paradox: What Do Love and Sex Have to Do with It?," *Frontiers in Ecology and Evolution* 6 (2018): 202; Ulrich H. Reichard and Christophe Boesch, eds., *Monogamy: Mating Strategies and Partnerships in Birds, Humans and Other Mammals* (Cambridge: Cambridge University Press, 2003).
5. Patricia Adair Gowaty, "Battles of the Sexes and Origins of Monogamy," in *Partnerships in Birds,* ed. Jeffrey M. Black (Oxford: Oxford University Press, 1996), 21–52; Kleiman, "Monogamy in Mammals"; Dixson, *Mammalian Sexuality;* Gray and Garcia, *Evolution and Human Sexual Behavior;* Carter and Perkeybile, "The Monogamy Paradox."
6. David M. Buss, "Human Mate Guarding," *Neuroendocrinology Letters* 23, no. 4 (2002): 23–29; Peter N. M. Brotherton and Petr E. Komers, "Mate Guarding and the Evolution of Social Monogamy in Mammals," in *Monogamy: Mating Strategies and Partnerships in Birds, Humans and Other Mammals,* ed. Ulrich H. Reichard and Christophe Boesch (Cambridge: Cambridge University Press, 2003), 42–58; Hanna Kokko and Lesley J. Morrell, "Mate Guarding, Male Attractiveness, and Paternity under Social Monogamy," *Behavioral Ecology* 16, no. 4 (2005): 724–31.
7. Samantha Leivers, Gillian Rhodes, and Leigh W. Simmons, "Sperm Competition in Humans: Mate Guarding Behavior Negatively Correlates with Ejaculate Quality," *PLOS One* 9, no. 9 (2014): e108099.
8. "Human Mate Guarding"; Todd K. Shackelford, Aaron T. Goetz, Faith E. Guta, and David P. Schmitt, "Mate Guarding and Frequent In-Pair Copulation in Humans: Concurrent or Compensatory Anti-Cuckoldry Tactics?," *Human Nature* 17 (2006): 239–52; David M. Buss, *The Dangerous Passion: Why Jealousy Is as Necessary as Love and Sex* (New York: Free Press, 2000).
9. Joyce H. Poole, "Mate Guarding, Reproductive Success and Female Choice in African Elephants," *Animal Behaviour* 37 (1989): 842–49.
10. Buss, *The Dangerous Passion.*
11. Mons Bendixen, Leif Edward Ottesen Kennair, and Trond Viggo Grøntvedt, "Forgiving the Unforgivable: Couples' Forgiveness and Expected Forgiveness of Emotional and Sexual Infidelity from an Error Management Theory Perspective," *Evolutionary Behavioral Sciences* 12, no. 4 (2018): 322–35.
12. Donald A. Dewsbury, "Effects of Novelty of Copulatory Behavior: The Coolidge Effect and Related Phenomena," *Psychological Bulletin* 89, no. 3 (1981): 464–82; L. A. Jordan and R. C. Brooks, "The Lifetime Costs of Increased Male Reproductive Effort: Courtship, Copulation, and the Coolidge Effect," *Journal of Evolutionary Biology* 23, no. 11 (2010): 2403–9; Susan M. Hughes, Toe Aung, Marissa A. Harrison, Jack N. LaFayette, and Gordon G. Gallup Jr., "Experimental Evidence for Sex Differences in Sexual Variety Preferences: Support for the Coolidge Effect in

Humans," *Archives of Sexual Behavior* 50 (2021): 495–509; Elisa Ventura-Aquino, Alonso Fernández-Guasti, and Raúl G. Paredes, "Hormones and the Coolidge Effect," *Molecular and Cellular Endocrinology* 467 (2018): 42–48; Garvin Vance and Todd K. Shackelford, "The Coolidge Effect," in *Encyclopedia of Animal Cognition and Behavior,* ed. Jennifer Vonk and Todd K. Shackelford (Cham, Switzerland: Springer, 2022), 1680–81.

13. Ventura-Aquino et al., "Hormones and the Coolidge Effect"; Dewsbury, "Effects of Novelty of Copulatory Behavior"; José L. Tlachi-López, Jose R. Eguibar, Alonso Fernández-Guasti, and Rosa Angélico Lucio, "Copulation and Ejaculation in Male Rats Under Sexual Satiety and the Coolidge Effect," *Physiology and Behavior* 106, no. 5 (2012): 626–30.

14. R. J. Levin, "Revisiting Post-Ejaculation Refractory Time — What We Know and What We Do Not Know in Males and in Females," *Journal of Sexual Medicine* 6, no. 9 (2009): 2376–89; William H. Masters and Virginia E. Johnson, *Human Sexual Response* (Boston: Little, Brown, 1966).

15. Hughes et al., "Experimental Evidence for Sex Differences in Sexual Variety Preferences"; Ísis Gomes Vasconcelos, "Coolidge Effect," in *Encyclopedia of Sexual Psychology and Behavior,* ed. Todd K. Shackelford (Cham, Switzerland: Springer, 2022).

16. Samantha J. Dawson, Kelly D. Suschinsky, and Martin L. Lalumière, "Habituation of Sexual Responses in Men and Women: A Test of the Preparation Hypothesis of Women's Genital Responses," *Journal of Sexual Medicine* 10, no. 4 (2013): 990–1000; Ellen Laan and Walter Everaerd, "Habituation of Female Sexual Arousal to Slides and Film," *Archives of Sexual Behavior* 24, no. 5 (1995): 517–41; Eric Koukounas and Ray Over, "Habituation of Male Sexual Arousal: Effects of Attentional Focus," *Biological Psychology* 58, no. 1 (2001): 49–64; Ray Over and Eric Koukounas, "Habituation of Sexual Arousal: Product and Process," *Annual Review of Sex Research* 6, no. 1 (1995): 187–223.

17. D. P. Schmitt and the International Sexuality Description Project, "Universal Sex Differences in the Desire for Sexual Variety: Tests from 52 Nations, 6 Continents, and 13 Islands," *Journal of Personality and Social Psychology* 85, no. 1 (2003): 85–104.

18. Schmitt et al., "Universal Sex Differences in the Desire for Sexual Variety"; Bruce M. King, "The Influence of Social Desirability on Sexual Behavior Surveys: A Review," *Archives of Sexual Behavior* 51, no. 3 (2022): 1495–501; Bente Træen, Nantje Fischer, and Ingela Lundin Kvalem, "Sexual Variety in Norwegian Men and Women of Different Sexual Orientations and Ages," *Journal of Sex Research* 59, no. 2 (2022): 238–47; Jes L. Matsick, Mary Kruk, Terri D. Conley, Amy C. Moors, and Ali Ziegler, "Gender Similarities and Differences in Casual Sex Acceptance Among Lesbian Women and Gay Men," *Archives of Sexual Behavior* 50, no. 3 (2021): 1151–66; Janet Shibley Hyde, "The Gender Similarities Hypothesis," *American Psychologist* 60, no. 6 (2005): 581–92.

19. Hyde, "The Gender Similarities Hypothesis."

20. Terri D. Conley, Amanda Ziegler, and Amy C. Moors, "Backlash from the Bedroom: Stigma Mediates Gender Differences in Acceptance of Casual Sex Offers," *Psychology of Women Quarterly* 37, no. 3 (2013): 392–407; Terri D. Conley, Justin D. Rubin, Jennifer L. Matsick, Amanda Ziegler, and Amy C. Moors, "Proposer Gender, Pleasure, and Danger in Casual Sex Offers Among Bisexual Women and Men," *Journal of Experimental*

Social Psychology 55 (2014): 80–88; Terri D. Conley, "Perceived Proposer Personality Characteristics and Gender Differences in Acceptance of Casual Sex Offers," *Journal of Personality and Social Psychology* 100, no. 2 (2011): 309–29.

21. "US Single People Under 50 Having Less Sex Since Roe Overturned, Study Finds," *The Guardian*, January 24, 2024.
22. Jankowiak and Hardgrave, "Individual and Societal Response to Sexual Betrayal"; William Jankowiak, Melissa D. Nell, and Amy Buckmaster, "Managing Infidelity: A Cross-Cultural Perspective," *Ethnology* 41, no. 1 (2002): 85–101.
23. Arezou Haseli, Mohammad Shariati, Amir Mohammad Nazari, Abbas Keramat, and Mohammad Hassan Emamian, "Infidelity and Its Associated Factors: A Systematic Review," *Journal of Sexual Medicine* 16, no. 8 (2019): 1155–69; Frank D. Fincham and Ryan W. May, "Infidelity in Romantic Relationships," *Current Opinion in Psychology* 13 (2017): 70–74; Adrian J. Blow and Kelly Hartnett, "Infidelity in Committed Relationships II: A Substantive Review," *Journal of Marital and Family Therapy* 31, no. 2 (2005): 217–33; Adrian J. Blow and Kelly Hartnett, "Infidelity in Committed Relationships I: A Methodological Review," *Journal of Marital and Family Therapy* 31, no. 2 (2005): 183–216; Alice Vossler, "Internet Infidelity 10 Years On: A Critical Review of the Literature," *Family Journal* 24, no. 4 (2016): 359–66; Richard D. McAnulty and Jennifer M. Brineman, "Infidelity in Dating Relationships," *Annual Review of Sex Research* 18, no. 1 (2007): 94–114; Ioannis Tsapelas, Helen E. Fisher, and Arthur Aron, "Infidelity: When, Where, Why," in *The Dark Side of Close Relationships II*, ed. William R. Cupach and Brian H. Spitzberg (New York: Routledge, 2010), 195–216; Tiffany DeLecce and Todd K. Shackelford, eds., *The Oxford Handbook of Infidelity* (Oxford: Oxford University Press, 2022); David C. Atkins, Donald H. Baucom, and Neil S. Jacobson, "Understanding Infidelity: Correlates in a National Random Sample," *Journal of Family Psychology* 15, no. 4 (2001): 735–49; Beth Warach, Robert F. Bornstein, Barbara S. Gorman, and Anne Moyer, "The Current State of Affairs in Infidelity Research: A Systematic Review and Meta-Analysis of Romantic Infidelity Prevalence and Its Moderators," *Personal Relationships* 31, no. 4 (2024): 1001–26.
24. Justin R. Garcia, James MacKillop, Elissa L. Aller, Ann M. Merriwether, David S. Wilson, and J. Koji Lum, "Associations Between Dopamine D4 Receptor Gene Variation with Both Infidelity and Sexual Promiscuity," *PLOS One* 5, no. 11 (2010): e14162.
25. Todd F. Heatherton, Lynn T. Kozlowski, Richard C. Frecker, and Karl O. Fagerström, "The Fagerström Test for Nicotine Dependence: A Revision of the Fagerström Tolerance Questionnaire," *British Journal of Addiction* 86, no. 9 (1991): 1119–27.
26. K. N. Kirby, Nancy M. Petry, and Warren K. Bickel, "Heroin Addicts Have Higher Discount Rates for Delayed Rewards than Non-Drug-Using Controls," *Journal of Experimental Psychology: General* 128, no. 1 (1999): 78–87.
27. Lars Versen, Sverre Iversen, Simon Dunnett, and Anders Björklund, eds., *Dopamine Handbook* (Oxford: Oxford University Press, 2009); K. M. Costa and Geoffrey Schoenbaum, "Dopamine," *Current Biology* 32, no. 15 (2022): R817–24; Wolfgang Schultz, "Dopamine Signals for Reward Value and Risk: Basic and Recent Data," *Behavioral and Brain Functions* 6 (2010): 1–9; Marc Fakhoury, ed., *The Brain Reward System* (New York: Humana Press, 2021); Judith Stellar, *The Neurobiology of Motivation and*

Reward (New York: Springer, 2012); Wolfgang Schultz, "Reward Signaling by Dopamine Neurons," *The Neuroscientist* 7, no. 4 (2001): 293–302.

28. Mark Jobling and Chris Tyler-Smith, *Human Evolutionary Genetics: Origins, Peoples and Disease* (New York: Garland, 2019); Ricki Lewis, *Human Genetics: Concepts and Applications,* 14th ed. (New York: McGraw-Hill, 2016); Gunnhild Saetre and Martin Ravinet, *Evolutionary Genetics: Concepts, Analysis, and Practice* (Oxford: Oxford University Press, 2019).

29. Garcia et al., "Associations Between Dopamine D4 Receptor Gene Variation with Both Infidelity and Sexual Promiscuity."

30. Benjamin C. Campbell, Anna Dreber, Coren L. Apicella, Dan T. A. Eisenberg, Peter B. Gray, Anthony C. Little, Justin R. Garcia, Richard S. Zamore, and J. Koji Lum, "Testosterone Exposure, Dopaminergic Reward, and Sensation-Seeking in Young Men," *Physiology and Behavior* 99, no. 4 (2010): 451–56; Anna Dreber, Coren L. Apicella, Dan T. A. Eisenberg, Justin R. Garcia, Richard S. Zamore, J. Koji Lum, and Benjamin C. Campbell, "The 7R Polymorphism in the Dopamine Receptor D4 Gene (DRD4) Is Associated with Financial Risk-Taking in Men," *Evolution and Human Behavior* 30, no. 2 (2009): 85–92; Jeffrey P. Carpenter, Justin R. Garcia, and J. Koji Lum, "Dopamine Receptor Genes Predict Risk Preferences, Time Preferences, and Related Economic Choices," *Journal of Risk and Uncertainty* 42, no. 3 (2011): 233–61; Anna Dreber, David G. Rand, Nina Wernerfelt, Justin R. Garcia, Maria G. Vilar, J. Koji Lum, and Richard J. Zeckhauser, "Dopamine and Risk Choices in Different Domains: Findings Among Serious Tournament Bridge Players," *Journal of Risk and Uncertainty* 43, no. 1 (2011): 19–38; Garcia et al., "Associations Between Dopamine D4 Receptor Gene Variation with Both Infidelity and Sexual Promiscuity"; John McGeary, "The DRD4 Exon 3 VNTR Polymorphism and Addiction-Related Phenotypes: A Review," *Pharmacology Biochemistry and Behavior* 93, no. 3 (2009): 222–29; Marcus R. Munafò, Burçin Yalcin, Simon A. Willis-Owen, and Jonathan Flint, "Association of the Dopamine D4 Receptor (DRD4) Gene and Approach-Related Personality Traits: Meta-Analysis and New Data," *Biological Psychiatry* 63, no. 2 (2008): 197–206; Avraham N. Kluger, Zahava Siegfried, and Richard P. Ebstein, "A Meta-Analysis of the Association Between DRD4 Polymorphism and Novelty Seeking," *Molecular Psychiatry* 7, no. 7 (2002): 712–17.

31. Lee Alan Dugatkin and Marie S. Alfieri, "Boldness, Behavioral Inhibition and Learning," *Ethology Ecology and Evolution* 15, no. 1 (2003): 43–49; Daniel J. Kruger, X. T. Wang, and Andreas Wilke, "Towards the Development of an Evolutionarily Valid Domain-Specific Risk-Taking Scale," *Evolutionary Psychology* 5, no. 3 (2007): 555–68; X. T. Wang, Daniel J. Kruger, and Andreas Wilke, "Life History Variables and Risk-Taking Propensity," *Evolution and Human Behavior* 30, no. 2 (2009): 77–84; Anders P. Møller and László Z. Garamszegi, "Between Individual Variation in Risk-Taking Behavior and Its Life History Consequences," *Behavioral Ecology* 23, no. 4 (2012): 843–53; David Sloan Wilson, Anne B. Clark, Katherine Coleman, and Theresa Dearstyne, "Shyness and Boldness in Humans and Other Animals," *Trends in Ecology and Evolution* 9, no. 11 (1994): 442–46; Andrew Sih, Alison Bell, and John C. Johnson, "Behavioral Syndromes: An Ecological and Evolutionary Overview," *Trends in Ecology and Evolution* 19, no. 7 (2004): 372–78.

32. Stein Engen, Jonathan Wright, Yimen G. Araya-Ajoy, and Bernt-Erik Sæther, "Phenotypic Evolution in Stochastic Environments: The Contribution of Frequency- and Density-Dependent Selection," *Evolution* 74, no. 9 (2020): 1923–41; Wilson et al., "Shyness and Boldness in Humans and Other Animals"; Sih et al., "Behavioral Syndromes."
33. Renée A. Duckworth and Alexander V. Badyaev, "Coupling of Dispersal and Aggression Facilitates the Rapid Range Expansion of a Passerine Bird," *Proceedings of the National Academy of Sciences* 104 (2007): 15017–22; Renée A. Duckworth and Keith W. Sockman, "Proximate Mechanisms of Behavioural Inflexibility: Implications for the Evolution of Personality Traits," *Functional Ecology* 26, no. 3 (2012): 559–66; Renée A. Duckworth, Alison L. Potticary, and Alexander V. Badyaev, "On the Origins of Adaptive Behavioral Complexity: Developmental Channeling of Structural Trade-Offs," in *Advances in the Study of Behavior*, vol. 50, ed. Marc Naguib, Louise Barrett, Susan D. Healy, Jeffrey Podos, Leigh W. Simmons, and Marlene Zuk (San Diego: Academic Press, 2018), 1–36; Angela N. Albers, Jared A. Jones, and Lynn Siefferman, "Behavioral Differences Among Eastern Bluebird Populations Could Be a Consequence of Tree Swallow Presence: A Pilot Study," *Frontiers in Ecology and Evolution* 5 (2017): 116.
34. Chun Chen, Linda Burton, Eliza Greenberger, and Julia Dmitrieva, "Population Migration and the Variation of Dopamine D4 Receptor (DRD4) Allele Frequencies Around the Globe," *Evolution and Human Behavior* 20, no. 5 (1999): 309–24; Luke J. Matthews and Paul M. Butler, "Novelty-Seeking DRD4 Polymorphisms Are Associated with Human Migration Distance Out-of-Africa After Controlling for Neutral Population Gene Structure," *American Journal of Physical Anthropology* 145, no. 3 (2011): 382–89; E. Wang, Y. C. Ding, P. Flodman, J. R. Kidd, K. K. Kidd, D. L. Grady, and Robert Moyzis, "The Genetic Architecture of Selection at the Human Dopamine Receptor D4 (DRD4) Gene Locus," *American Journal of Human Genetics* 74, no. 5 (2004): 931–44.
35. G. L. Hunt Jr. and M. W. Hunt, "Female-Female Pairing in Western Gulls (*Larus occidentalis*) in Southern California," *Science* 196, no. 4297 (1977): 1466–67.
36. Paul L. Vasey, "Same-Sex Sexual Partner Preference in Hormonally and Neurologically Unmanipulated Animals," *Annual Review of Sex Research* 13, no. 1 (2002): 141–79; Nathan W. Bailey and Marlene Zuk, "Same-Sex Sexual Behavior and Evolution," *Trends in Ecology and Evolution* 24, no. 8 (2009): 439–46; Bruce Bagemihl, *Biological Exuberance: Animal Homosexuality and Natural Diversity* (New York: St. Martin's Press, 1999).
37. Robin R. Baker and Mark A. Bellis, *Human Sperm Competition: Copulation, Masturbation, and Infidelity* (London: Chapman and Hall, 1995).
38. Kermyt Anderson, "How Well Does Paternity Confidence Match Actual Paternity? Evidence from Worldwide Nonpaternity Rates," *Current Anthropology* 47, no. 3 (2006): 513–20.
39. Baker and Bellis, *Human Sperm Competition;* Anderson, "How Well Does Paternity Confidence Match Actual Paternity?"
40. Dylan Selterman, Justin R. Garcia, and Ioannis Tsapelas, "What Do People Do, Say, and Feel When They Have Affairs? Associations Between Extradyadic Infidelity Motives with Behavioral, Emotional, and Sexual Outcomes," *Journal of Sex and Marital*

Therapy 47, no. 3 (2021): 238–52; Todd K. Shackelford, "Divorce as a Consequence of Spousal Infidelity," in *Romantic Love and Sexual Behavior*, ed. Victor de Munck (New York: Praeger, 1998), 135–53; Emily Q. Hoy and Vivian Y. Oh, "The Consequences of Spousal Infidelity for Long-Term Chronic Health: A Two-Wave Longitudinal Analysis," *Journal of Social and Personal Relationships* 41, no. 12 (2024): 3720–40; Paul E. Mullen, "The Crime of Passion and the Changing Cultural Construction of Jealousy," *Criminal Behaviour and Mental Health* 3, no. 1 (1993): 1–11; Adrienne Howe, *Crimes of Passion Since Shakespeare: Red Mist Rage Unmasked* (London: Routledge, 2023); Hila Dayan, "Femicide and the 'Heat of Passion' Criminal Doctrine," in *The Routledge International Handbook on Femicide and Feminicide*, ed. Myrna Dawson and Saide Mobayed Vega (London: Routledge, 2023), 443–52; Buss, *The Dangerous Passion*; Daniel J. Kruger, Maryanne L. Fisher, and Christine J. Fitzgerald, "Factors Influencing the Intended Likelihood of Exposing Sexual Infidelity," *Archives of Sexual Behavior* 44 (2015): 1697–704; Maryanne L. Fisher and Amanda Tiller, "Cues to Infidelity," in *Encyclopedia of Evolutionary Psychological Science*, ed. Todd K. Shackelford and Viviana A. Weekes-Shackelford (Cham, Switzerland: Springer, 2021), 1631–38.

41. Gray and Garcia, *Evolution and Human Sexual Behavior*; Helen E. Fisher, "Serial Monogamy and Clandestine Adultery: Evolution and Consequences of the Dual Human Reproductive Strategy," in *Applied Evolutionary Psychology*, ed. S. Craig Roberts (Oxford: Oxford University Press, 2012), 93–111; Steven W. Gangestad, "Evidence for Adaptations for Female Extra-Pair Mating in Humans: Thoughts on Current Status and Future Directions," in *Female Infidelity and Paternal Uncertainty: Evolutionary Perspectives on Male Anti-Cuckoldry Tactics*, ed. Steven M. Platek and Todd K. Shackelford (New York: Cambridge University Press, 2006), 37–57; Tsapelas et al., "Infidelity"; Buss, *The Dangerous Passion*; Selterman et al., "What Do People Do, Say, and Feel When They Have Affairs?"; Esther Perel, *The State of Affairs: Rethinking Infidelity* (New York: HarperCollins, 2017).

42. Justin R. Garcia, Chris Reiber, Sean G. Massey, and Ann M. Merriwether, "Sexual Hookup Culture: A Review," *Review of General Psychology* 16, no. 2 (2012): 161–76; Justin R. Garcia and Chris Reiber, "Hook-Up Behavior: A Biopsychosocial Perspective," *Journal of Social, Evolutionary, and Cultural Psychology* 2, no. 4 (2008): 192–208.

43. Dylan Selterman, Justin R. Garcia, and Ioannis Tsapelas, "Motivations for Extradyadic Infidelity Revisited," *Journal of Sex Research* 56, no. 3 (2019): 273–86.

44. Dylan Selterman, Justin R. Garcia, and Irene Tsapelas, "What do people do, say, and feel when they have affairs? Associations between extradyadic infidelity motives with behavioral, emotional, and sexual outcomes," *Journal of Sex & Marital Therapy* 47, no. 3 (2021): 238-252.

45. David M. Buss, "Sexual and Emotional Infidelity: Evolved Gender Differences in Jealousy Prove Robust and Replicable," *Perspectives on Psychological Science* 13, no. 2 (2018): 155–60; David M. Buss, Randy J. Larsen, Drew Westen, and Jennifer Semmelroth, "Sex Differences in Jealousy: Evolution, Physiology, and Psychology," *Psychological Science* 3, no. 4 (1992): 251–56; Todd K. Shackelford, David M. Buss, and Krista Bennett, "Forgiveness or Breakup: Sex Differences in Responses to a Partner's Infidelity," *Cognition and Emotion* 16, no. 2 (2002): 299–307; Menelaos Apostolou, Christina Constantinou, and Angela Zalaf, "How People React to Their Partners' Infidelity: An Explorative

Study," *Personal Relationships* 29, no. 4 (2022): 913–32; Menelaos Apostolou, Andria Aristidou, and Charis Eraclide, "Reactions to and Forgiveness of Infidelity: Exploring Severity, Length of Relationship, Sex, and Previous Experience Effects," *Adaptive Human Behavior and Physiology* 5 (2019): 317–30; Amanda E. Guitar, Glenn Geher, Daniel J. Kruger, Justin R. Garcia, Maryanne L. Fisher, and Christine J. Fitzgerald, "Defining and Interpreting Definitions of Emotional and Sexual Infidelity," *Current Psychology* 36, no. 3 (2017): 434–46; Daniel J. Kruger, Maryanne L. Fisher, Christine J. Fitzgerald, Justin R. Garcia, Glenn Geher, and Amanda E. Guitar, "Sexual and Emotional Aspects Are Distinct Components of Infidelity and Unique Predictors of Anticipated Distress," *Evolutionary Psychological Sciences* 1, no. 1 (2015): 44–51; Todd K. Shackelford, Gregory J. LeBlanc, and Erik Drass, "Emotional Reactions to Infidelity," *Cognition and Emotion* 14, no. 5 (2000): 643–59; Christine R. Harris, "Psychophysiological Responses to Imagined Infidelity: The Specific Innate Modular View of Jealousy Reconsidered," *Journal of Personality and Social Psychology* 78, no. 6 (2000): 1082–91; Christine R. Harris, "Factors Associated with Jealousy over Real and Imagined Infidelity: An Examination of the Social-Cognitive and Evolutionary Psychology Perspectives," *Psychology of Women Quarterly* 27, no. 4 (2003): 319–29; Christine R. Harris, "A Review of Sex Differences in Sexual Jealousy, Including Self-Report Data, Psychophysiological Responses, Interpersonal Violence, and Morbid Jealousy," *Personality and Social Psychology Review* 7, no. 2 (2003): 102–28; Christine R. Harris and Nicholas Christenfeld, "Jealousy and Rational Responses to Infidelity Across Gender and Culture," *Psychological Science* 7, no. 6 (1996): 378–79; Buss, *The Dangerous Passion*.
46. Buss et al., "Sex Differences in Jealousy."
47. Bram P. Buunk, Andreas Angleitner, Volker Oubaid, and David M. Buss, "Sex Differences in Jealousy in Evolutionary and Cultural Perspective: Tests from the Netherlands, Germany, and the United States," *Psychological Science* 7 (1996): 359–63; Buss, *The Dangerous Passion*; Selterman et al., "What Do People Do, Say, and Feel When They Have Affairs? Associations Between Extradyadic Infidelity Motives with Behavioral, Emotional, and Sexual Outcomes," *Journal of Sex and Marital Therapy* 47, no. 3 (2021): 238–52; Shackelford et al., "Emotional Reactions to Infidelity"; Elizabeth Q. Hoy and V. Y. Oh, "The Consequences of Spousal Infidelity for Long-Term Chronic Health: A Two-Wave Longitudinal Analysis," *Journal of Social and Personal Relationships* 41, no. 12 (2024): 3720–40; Jennifer F. Landolfi, Geoffrey Geher, and Alice Andrews, "The Role of Stimulus Specificity on Infidelity Reactions: Seeing Is Disturbing," *Current Psychology* 26 (2007): 46–59.
48. John Gottman, Julie Gottman, and Mary A. McNulty, "The Role of Trust and Commitment in Love Relationships," in *Foundations for Couples' Therapy*, ed. Jennifer Fitzgerald (New York: Routledge, 2017), 438–52; John Gottman and Nan Silver, *What Makes Love Last?: How to Build Trust and Avoid Betrayal* (New York: Simon & Schuster, 2012); Perel, *The State of Affairs*; Jankowiak and Hardgrave, "Individual and Societal Response to Sexual Betrayal"; Jankowiak et al., "Managing Infidelity."
49. Justin K. Mogilski, Simon D. Reeve, Sylis C. A. Nicolas, Sarah H. Donaldson, Virginia E. Mitchell, and Lisa L. M. Welling, "Jealousy, Consent, and Compersion Within Monogamous and Consensually Non-Monogamous Romantic Relationships," *Archives of Sexual Behavior* 48 (2019): 1811–28; Clara Andersson, "Drawing the Line at Infidelity: Negotiating Relationship Morality in a Swedish Context of Consensual

Non-Monogamy," *Journal of Social and Personal Relationships* 39, no. 7 (2022): 1917–33; Marissa A. Davala and Grace A. Mims, "Monogamy and Consensual Non-Monogamy Fidelity: Preventing Relationship Crisis," in *Infidelity: A Practitioner's Guide to Working with Couples in Crisis,* 2nd ed., ed. Paul R. Peluso and Taylor J. Irvine (New York: Routledge, 2024), 21–37; Forrest Hangen, Dev Crasta, and Ronald D. Rogge, "Delineating the Boundaries Between Nonmonogamy and Infidelity: Bringing Consent Back into Definitions of Consensual Nonmonogamy with Latent Profile Analysis," *Journal of Sex Research* 57, no. 4 (2020): 438–57; J. L. Stewart, Christopher B. Stults, and Annie Ristuccia, "Consensual Non-Monogamy Relationship Rules Among Young Gay and Bisexual Men: A Dyadic Qualitative Analysis," *Archives of Sexual Behavior* 50, no. 4 (2021): 1505–20.

50. Jankowiak et al., "Managing Infidelity: A Cross-Cultural Perspective."
51. Gayle Brewer, David Hunt, Georgia James, and Laura Abell, "Dark Triad Traits, Infidelity, and Romantic Revenge," *Personality and Individual Differences* 83 (2015): 122–27; Gayle Brewer, Anna Guothova, and Dimitrios Tsivilis, "'But It Wasn't Really Cheating': Dark Triad Traits and Perceptions of Infidelity," *Personality and Individual Differences* 202 (2023): 111987; Daniel N. Jones and David A. Weiser, "Differential Infidelity Patterns Among the Dark Triad," *Personality and Individual Differences* 57 (2014): 20–24.
52. Ken Yasukawa and William A. Searcy, "Aggression in Female Red-Winged Blackbirds: A Strategy to Ensure Male Parental Investment," *Behavioral Ecology and Sociobiology* 11 (1982): 13–17; Rachel Olendorf, Thomas Getty, Kim Scribner, and Scott K. Robinson, "Male Red-Winged Blackbirds Distrust Unreliable and Sexually Attractive Neighbours," *Proceedings of the Royal Society B: Biological Sciences* 271, no. 1543 (2004): 1033–38.
53. Perel, *The State of Affairs;* Gottman and Silver, *What Makes Love Last?*

Chapter Eight: Break

1. Amy Rauer, Allison K. Sabey, Christine M. Proulx, and Brenda L. Volling, "What Are the Marital Problems of Happy Couples? A Multimethod, Two-Sample Investigation," *Family Process* 59, no. 3 (2020): 1275–92.
2. Shelley Duncan, Judith van Hooff, and Jason Carter, "Living Together Apart: Size and Significance of Co-Residency Following Relationship Breakdown in Contemporary Britain," *Sociological Research Online* 30, no. 1 (2024): 78–95.
3. Tony Walter, *On Bereavement: The Culture of Grief* (Maidenhead, UK: McGraw-Hill Education, 1999); Robert A. Neimeyer, Darcy L. Harris, Howard R. Winokuer, and Gordon F. Thornton, eds., *Grief and Bereavement in Contemporary Society: Bridging Research and Practice* (New York: Routledge, 2011); Randolph M. Nesse, "An Evolutionary Framework for Understanding Grief," in *Spousal Bereavement in Late Life,* ed. Deborah Carr, Randolph M. Nesse, and Camille B. Wortman (New York: Springer, 2005), 195–226; Rosemary Gross, *The Psychology of Grief* (New York: Routledge, 2018); John Archer, *The Nature of Grief: The Evolution and Psychology of Reactions to Loss* (New York: Routledge, 1999).
4. Christine E. Morris and Cassandra Reiber, "Frequency, Intensity and Expression of Post-Relationship Grief," *EvoS Journal: The Journal of the Evolutionary Studies*

Consortium 3, no. 1 (2011): 1–11; Jessica M. DeGroot and Heather J. Carmack, "Accidental and Purposeful Triggers of Post-Relationship Grief," *Journal of Loss and Trauma* 28, no. 4 (2023): 391–403; Paul A. Boelen and Maurice A. Van den Hout, "Inclusion of Other in the Self and Breakup-Related Grief Following Relationship Dissolution," *Journal of Loss and Trauma* 15, no. 6 (2010): 534–47; Helen Fisher, "Broken Hearts: The Nature and Risks of Romantic Rejection," in *Romance and Sex in Adolescence and Emerging Adulthood,* ed. Ann C. Crouter and Alan Booth (New York: Psychology Press, 2014), 3–28.

5. Margaret S. Stroebe, Henk A. W. Schut, and Wolfgang Stroebe, "Health Consequences of Bereavement: A Review," *Lancet Infectious Diseases* 370, no. 9603 (2007): 1960–73; Tiffany Field, "Romantic Breakups, Heartbreak and Bereavement — Romantic Breakups," *Psychology* 2, no. 4 (2011): 382–87; Adela Mirsu-Paun and Jason A. Oliver, "How Much Does Love Really Hurt? A Meta-Analysis of the Association Between Romantic Relationship Quality, Breakups and Mental Health Outcomes in Adolescents and Young Adults," *Journal of Relationships Research* 8 (2017): e5; Michael J. Wilson, Kayla Mansour, Zac E. Seidler, John L. Oliffe, Simon M. Rice, Paul Sharp, Christopher J. Greenwood, and Jacqui A. Macdonald, "Intimate Partner Relationship Breakdown and Suicidal Ideation in a Large Representative Cohort of Australian Men," *Journal of Affective Disorders* 372 (2025): 618–26; Helen E. Fisher, Lucy L. Brown, Arthur Aron, Greg Strong, and Debra Mashek, "Reward, Addiction, and Emotion Regulation Systems Associated with Rejection in Love," *Journal of Neurophysiology* 104, no. 1 (2010): 51–60; Ana del Palacio-Gonzalez, David A. Clark, and Lucia F. O'Sullivan, "Distress Severity Following a Romantic Breakup Is Associated with Positive Relationship Memories Among Emerging Adults," *Emerging Adulthood* 5, no. 4 (2017): 259–67.

6. Morris and Reiber, "Frequency, Intensity and Expression of Post-Relationship Grief"; Mirsu-Paun and Oliver, "How Much Does Love Really Hurt?"; Tiffany Field, Samantha Poling, Shantay Mines, Miguel Diego, Debra Bendell, and Martha Pelaez, "Trauma Symptoms Following Romantic Breakups," *Journal of Psychology and Clinical Psychiatry* 12, no. 2 (2021): 37–42; Anne M. Verhallen, Remco J. Renken, Jan-Bernard C. Marsman, and Gert J. Ter Horst, "Romantic Relationship Breakup: An Experimental Model to Study Effects of Stress on Depression (-Like) Symptoms," *PLOS One* 14, no. 5 (2019): e0217320; Bhaveena Studley and Man Cheung Chung, "Posttraumatic Stress and Well-Being Following Relationship Dissolution: Coping, Posttraumatic Stress Disorder Symptoms from Past Trauma, and Traumatic Growth," *Journal of Loss and Trauma* 20, no. 4 (2015): 317–35.

7. Susan M. Monroe, Paul Rohde, John R. Seeley, and Peter M. Lewinsohn, "Life Events and Depression in Adolescence: Relationship Loss as a Prospective Risk Factor for First Onset of Major Depressive Disorder," *Journal of Abnormal Psychology* 108, no. 4 (1999): 606–14; Esther Ohenewa and Ansie Meyer-Weitz, "Suicide Risk in the Context of Relationship Breakup Among Young Adults: Emotional Pain and Mental Health Indicators," *Advances in Mental Health* (2024): 1–19.

8. Mirsu-Paun and Oliver, "How Much Does Love Really Hurt?"; Galena K. Rhoades, Claire M. Kamp Dush, David C. Atkins, Scott M. Stanley, and Howard J. Markman, "Breaking Up Is Hard to Do: The Impact of Unmarried Relationship Dissolution on Mental Health and Life Satisfaction," *Journal of Family Psychology* 25, no. 3 (2011):

366–74; Mark A. Whisman, Julie M. Salinger, and David A. Sbarra, "Relationship Dissolution and Psychopathology," *Current Opinion in Psychology* 43 (2022): 199–204; Fisher, "Broken Hearts."

9. Fisher et al., "Reward, Addiction, and Emotion Regulation Systems Associated with Rejection in Love."

10. Dawn M. Carpenter, Emily A. Elstad, Amanda J. Sage, Laura L. Geryk, Robert F. DeVellis, and Susan J. Blalock, "The Relationship Between Partner Information-Seeking, Information-Sharing, and Patient Medication Adherence," *Patient Education and Counseling* 98, no. 1 (2015): 120–24; William J. Doherty, Henry G. Schrott, Linda Metcalf, and Linda Iasiello-Vailas, "Effect of Spouse Support and Health Beliefs on Medication Adherence," *Journal of Family Practice* 17, no. 5 (1983): 837–41; Mary A. P. Stephens, Erin M. Fekete, Melissa M. Franks, Karen S. Rook, Jennifer A. Druley, and Kathryn Greene, "Spouses' Use of Pressure and Persuasion to Promote Osteoarthritis Patients' Medical Adherence After Orthopedic Surgery," *Health Psychology* 28, no. 1 (2009): 48–55; Misook L. Chung, Debra K. Moser, Terry A. Lennie, and Barbara Riegel, "Spouses Enhance Medication Adherence in Patients with Heart Failure," *Circulation* 114, suppl. 18 (2006): II-518.

11. Sarah de Regt, Dimitri Mortelmans, and Tim Marynissen, "Financial Consequences of Relationship Dissolution: A Longitudinal Comparison of Formerly Married and Unmarried Cohabiting Men and Women," *Sociology* 47, no. 1 (2013): 90–108; Liana C. Sayer, "Economic Aspects of Divorce and Relationship Dissolution," in *Handbook of Divorce and Relationship Dissolution,* 2nd ed., ed. Mark A. Fine and John H. Harvey (New York: Psychology Press, 2013), 385–406.

12. Fisher et al., "Reward, Addiction, and Emotion Regulation Systems Associated with Rejection in Love."

13. Helen E. Fisher, "The Tyranny of Love: Love Addiction — An Anthropologist's View," in *Behavioral Addictions: Criteria, Evidence and Treatment,* ed. Kenneth P. Rosenberg and Laura Curtiss Feder (London: Elsevier/Academic Press, 2014), 237–60; Marina Bolshakova, Helen E. Fisher, Henri-Jean Aubin, and Steve Sussman, "Passionate Love Addiction: An Evolutionary Survival Mechanism That Can Go Terribly Wrong," in *Cambridge Handbook of Substance and Behavioral Addictions,* ed. Steve Sussman (Cambridge: Cambridge University Press, 2020), 262–70; Helen E. Fisher, Xiaomeng Xu, Arthur Aron, and Lucy L. Brown, "Intense, Passionate, Romantic Love: A Natural Addiction? How the Fields That Investigate Romance and Substance Abuse Can Inform Each Other," *Frontiers in Psychology* 7 (2016): 1903.

14. Fisher et al., "Reward, Addiction, and Emotion Regulation Systems Associated with Rejection in Love."

15. Tessa S. Soloway, Ruxandra Gica, and Matthew R. Langlais, "To Break or Not to Break? Understanding the Motivations and Consequences of Taking Breaks in Romantic Relationships," *Journal of Couple and Relationship Therapy* 24, no. 1 (2025): 1–23.

16. Morris and Reiber, "Frequency, Intensity and Expression of Post-Relationship Grief"; Fisher, "Broken Hearts"; Matthew Larson, Gary Sweeten, and Alex R. Piquero, "With or Without You? Contextualizing the Impact of Romantic Relationship Breakup on Crime Among Serious Adolescent Offenders," *Journal of Youth and Adolescence* 45 (2016): 54–72; Heather A. Love, David P. Nalbone, Lynn L. Hecker, Kayla A. Sweeney,

and Priyanka Dharnidharka, "Suicidal Risk Following the Termination of Romantic Relationships," *Crisis* 39, no. 3 (2017): 166–74; Poh Choo, Timothy Levine, and Elaine Hatfield, "Gender, Love Schemas, and Reactions to Romantic Break-Ups," *Journal of Social Behavior and Personality* 11, no. 5 (1996): 143–60.

17. A. S. Marcotte, Amanda N. Gesselman, Thomas A. Reynolds, and Justin R. Garcia, "Young Adults' Romantic Investment Behaviors on Social Media," *Personal Relationships* 28, no. 4 (2021): 822–39.
18. K. D. Coduto and A. McDonald, "'Delete It and Move On': Digital Management of Shared Sexual Content After a Breakup," in *Proceedings of the 2024 CHI Conference on Human Factors in Computing Systems* (2024): 918, 1–16.
19. Marcotte et al., "Young Adults' Romantic Investment Behaviors on Social Media."
20. Xiaomeng Xu, Lucy Brown, Arthur Aron, Guikang Cao, Tingyong Feng, Bianca Acevedo, and Xuchu Weng, "Regional Brain Activity During Early-Stage Intense Romantic Love Predicted Relationship Outcomes After 40 Months: An fMRI Assessment," *Neuroscience Letters* 526, no. 1 (2012): 33–38; Bianca P. Acevedo, "Neural Correlates of Human Attachment: Evidence from fMRI Studies of Adult Pair-Bonding," in *Bases of Adult Attachment: Linking Brain, Mind, and Behavior,* ed. Vivian Zayas and Cindy Hazan (New York: Academic Press, 2015), 185–94; Bianca P. Acevedo, Arthur Aron, Helen E. Fisher, and Lucy L. Brown, "Neural Correlates of Long-Term Intense Romantic Love," *Social Cognitive and Affective Neuroscience* 7, no. 2 (2012): 145–59; Bianca P. Acevedo, Arthur Aron, Helen E. Fisher, and Lucy L. Brown, "Neural Correlates of Marital Satisfaction and Well-Being: Reward, Empathy, and Affect," *Clinical Neuropsychiatry* 9, no. 1 (2012): 20–31.
21. René M. Dailey, *On-Again, Off-Again Relationships: Navigating (In)stability in Romantic Relationships* (Cambridge: Cambridge University Press, 2020); René M. Dailey, Borae Jin, Abigail Pfiester, and Gary Beck, "On-Again/Off-Again Dating Relationships: What Keeps Partners Coming Back?," *Journal of Social Psychology* 151, no. 4 (2011): 417–40.
22. Gianluca Zara and David Farrington, *Criminal Recidivism: Explanation, Prediction and Prevention* (New York: Routledge, 2016); Fisher et al., "Intense, Passionate, Romantic Love: A Natural Addiction?"
23. Susan Sprecher, "Two Sides to the Breakup of Dating Relationships," *Personal Relationships* 1, no. 3 (1994): 199–222; Michael J. Rosenfeld, "Who Wants the Breakup? Gender and Breakup in Heterosexual Couples," in *Social Networks and the Life Course: Integrating the Development of Human Lives and Social Relational Networks,* ed. Duane F. Alwin, Diane H. Felmlee, and Derek A. Kreager (Cham, Switzerland: Springer, 2018), 221–43.
24. Rosenfeld, "Who Wants the Breakup?"
25. Paul J. Ceglarek, Lisa A. Darbes, Rob Stephenson, and José A. Bauermeister, "Breakup-Related Appraisals and the Psychological Well-Being of Young Adult Gay and Bisexual Men," *Journal of Gay and Lesbian Mental Health* 21, no. 3 (2017): 256–74; John M. Gottman, Robert W. Levenson, James Gross, Barbara L. Frederickson, Kim McCoy, Leah Rosenthal, Anna Ruef, and Dan Yoshimoto, "Correlates of Gay and Lesbian Couples' Relationship Satisfaction and Relationship Dissolution," *Journal of Homosexuality* 45, no. 1 (2003): 23–43; Lawrence A. Kurdek, "The Dissolution of Gay and Lesbian Couples," *Journal of Social and Personal Relationships* 8, no. 2 (1991): 265–78.

26. J.K. Mogilski, S.D. Reeve, S.C. Nicolas, S.H. Donaldson, V.E. Mitchell, and L.L. Welling, "Jealousy, Consent, and Compersion Within Monogamous and Consensually Non-Monogamous Romantic Relationships," *Archives of Sexual Behavior* 48 (2019): 1811–28.
27. Matthew D. Johnson, Justin A. Lavner, Scott M. Stanley, and Galena K. Rhoades, "Gender Differences — or the Lack Thereof — in the Prediction of Relationship Dissolution Among Unmarried Mixed-Gender Couples from the United States," *Journal of Social and Personal Relationships* 41, no. 11 (2024): 3316–36.
28. Iona Abrahamson, Rafat Hussain, Adeel Khan, and Margot J. Schofield, "What Helps Couples Rebuild Their Relationship After Infidelity?," *Journal of Family Issues* 33, no. 11 (2012): 1494–519; Stephen T. Fife, Jacob D. Gossner, Alex Theobald, Emma Allen, Ariana Rivero, and Heather Koehl, "Couple Healing from Infidelity: A Grounded Theory Study," *Journal of Social and Personal Relationships* 40, no. 12 (2023): 3882–905.
29. John Gottman, Julie Gottman, and Mary A. McNulty, "The Role of Trust and Commitment in Love Relationships," in *Foundations for Couples' Therapy*, ed. Jennifer Fitzgerald (New York: Routledge, 2017), 438–52; John Gottman and Nan Silver, *What Makes Love Last? How to Build Trust and Avoid Betrayal* (New York: Simon & Schuster, 2012); Carlos Perez, Stephen T. Fife, Dane Eggleston, and Jason B. Whiting, "Justifying by Degrees: A Grounded Theory of Men's Decision-Making Process in Infidelity," *Journal of Marital and Family Therapy* 49, no. 4 (2023): 879–98; Cassandra Alexopoulos, "Justify My Love: Cognitive Dissonance Reduction Among Perpetrators of Online and Offline Infidelity," *Journal of Social and Personal Relationships* 38, no. 12 (2021): 3669–91; Nicolle Marie Zapien, "Decision Science, Risk Perception, and Infidelity," *SAGE Open* 7, no. 1 (2017).
30. Christopher R. Agnew, Benjamin W. Hadden, and Kenneth Tan, "It's About Time: Readiness, Commitment, and Stability in Close Relationships," *Social Psychological and Personality Science* 10, no. 8 (2019): 1046–55; Benjamin W. Hadden and Christopher R. Agnew, "Commitment Readiness: Timing, the Self, and Close Relationships," in *Interpersonal Relationships and the Self-Concept*, ed. Brent A. Mattingly, Kevin P. McIntyre, and Gary W. Lewandowski (New York: Routledge, 2020), 53–67; Ximena B. Arriaga, Lucy L. Hunt, and Christopher R. Agnew, "The Time Has Come: Integrating Time and Timing into Our Understanding of Close Relationships," *Psychological Inquiry* 30, no. 1 (2019): 34–38.
31. Corinne F. Belu and Lucia F. O'Sullivan, "It's Just a Little Crush: Attraction to an Alternative and Romantic Relationship Quality, Breakups and Infidelity," *Journal of Sex Research* (2025): 1–14; Sarah A. Vannier and Lucia F. O'Sullivan, "Great Expectations: Examining Unmet Romantic Expectations and Dating Relationship Outcomes Using an Investment Model Framework," *Journal of Social and Personal Relationships* 35, no. 8 (2018): 1045–66; Brenda H. Lee and Lucia F. O'Sullivan, "Walk the Line: How Successful Are Efforts to Maintain Monogamy in Intimate Relationships?" *Archives of Sexual Behavior* 48 (2019): 1735–48; Corinne F. Belu and Lucia F. O'Sullivan, "Roving Eyes: Predictors of Crushes in Ongoing Romantic Relationships and Implications for Relationship Quality," *Journal of Relationships Research* 10 (2019): e2; Brenda H. Lee and Lucia F. O'Sullivan, "Ain't Misbehavin? Monogamy Maintenance Strategies in Heterosexual Romantic Relationships," *Personal Relationships* 25, no. 2 (2018): 205–32; Kirstian A. V. Gibson, Ashley E. Thompson, and Lucia F.

O'Sullivan, "Love Thy Neighbour: Personality Traits, Relationship Quality, and Attraction to Others as Predictors of Infidelity Among Young Adults," *Canadian Journal of Human Sexuality* 25, no. 3 (2016): 186–98.
32. Annamieke J. M. Van den Tol and Jane Edwards, "Listening to Sad Music in Adverse Situations: How Music Selection Strategies Relate to Self-Regulatory Goals, Listening Effects, and Mood Enhancement," *Psychology of Music* 43, no. 4 (2014): 473–94; Kaitlin Woolley and Ayelet Fishbach, "Motivating Personal Growth by Seeking Discomfort," *Psychological Science* 33, no. 4 (2022): 510–23; Brennan McDonald, Anne Böckler, and Philipp Kanske, "Soundtrack to the Social World: Emotional Music Enhances Empathy, Compassion, and Prosocial Decisions but Not Theory of Mind," *Emotion* 22, no. 1 (2022): 19–29.
33. Liila Taruffi and Stefan Koelsch, "The Paradox of Music-Evoked Sadness: An Online Survey," *PLOS One* 9, no. 10 (2014): e110490.

Chapter Nine: Care

1. John S. Rolland, "In Sickness and in Health: The Impact of Illness on Couples' Relationships," *Journal of Marital and Family Therapy* 20, no. 4 (1994): 327–47; Janice K. Kiecolt-Glaser and Stephanie J. Wilson, "Lovesick: How Couples' Relationships Influence Health," *Annual Review of Clinical Psychology* 13, no. 1 (2017): 421–43; George C. Karantzas et al., "Dealing with Loss in the Face of Disasters and Crises: Integrating Interpersonal Theories of Couple Adaptation and Functioning," *Current Opinion in Psychology* 43 (2022): 129–38; Mylène Hasdenteufel and Bruno Quintard, "Dyadic Experiences and Psychosocial Management of Couples Facing Advanced Cancer: A Systematic Review of the Literature," *Frontiers in Psychology* 13 (2022): 827947; Lauren J. Trump and Timothy J. Mendenhall, "Couples Coping with Cardiovascular Disease: A Systematic Review," *Families, Systems, and Health* 35, no. 1 (2017): 58–69; Dohye Kang et al., "Divorce After Breast Cancer Diagnosis and Its Impact on Quality of Life," *Palliative and Supportive Care* 20, no. 6 (2022): 807–16.
2. Lori E. Miller, "Sources of Uncertainty in Cancer Survivorship," *Journal of Cancer Survivorship* 6 (2012): 431–40.
3. Karen Kayser, Linda E. Watson, and Janine T. Andrade, "Cancer as a 'We-Disease': Examining the Process of Coping from a Relational Perspective," *Families, Systems, and Health* 25, no. 4 (2007): 404–18.
4. Atle Dyregrov and Rolf Gjestad, "Losing a Child: The Impact on Parental Sexual Activity," *Bereavement Care* 31, no. 1 (2012): 18–24; Sara Albuquerque, Marco Pereira, and Isabel Narciso, "Couple's Relationship After the Death of a Child: A Systematic Review," *Journal of Child and Family Studies* 25 (2016): 30–53; Atle Dyregrov and Kari Dyregrov, "Parents' Perception of Their Relationship Following the Loss of a Child," *OMEGA–Journal of Death and Dying* 76, no. 1 (2017): 35–52; Atle Dyregrov, Rolf Gjestad, and Kari Dyregrov, "Parental Relationships Following the Loss of a Child," *Journal of Loss and Trauma* 25, no. 3 (2020): 224–44; Anna Salakari, Marja Kaunonen, and Auli A. L. Aho, "Negative Changes in a Couple's Relationship After a Child's Death," *Interpersona: An International Journal on Personal Relationships* 8, no. 2 (2014): 193–209.

5. Annelies K. Hagemeister and Paul C. Rosenblatt, "Grief and the Sexual Relationship of Couples Who Have Experienced a Child's Death," *Death Studies* 21, no. 3 (1997): 231–52.
6. Buket Nalbant, Antonia Karger, and Thomas Zimmermann, "Cancer and Relationship Dissolution: Perspective of Partners of Cancer Patients," *Frontiers in Psychology* 12 (2021): 624902; Adrian Furnham, "Belief in a Just World: Research Progress over the Past Decade," *Personality and Individual Differences* 34, no. 5 (2003): 795–817; Laura T. Matthews and Samuel J. Marwit, "Examining the Assumptive World Views of Parents Bereaved by Accident, Murder, and Illness," *OMEGA—Journal of Death and Dying* 48, no. 2 (2004): 115–36; Albuquerque et al., "Couple's Relationship After the Death of a Child"; Luis E. Oliver, "Effects of a Child's Death on the Marital Relationship: A Review," *OMEGA—Journal of Death and Dying* 39, no. 3 (1999): 197–227; Jake M. Najman, John C. Vance, Fran Boyle, Gary Embleton, Bill Foster, and John Thearle, "The Impact of a Child Death on Marital Adjustment," *Social Science and Medicine* 37, no. 8 (1993): 1005–10.
7. Lerner, Melvin J. (1980). "The Belief in a Just World: A Fundamental Delusion." *Perspectives in Social Psychology*. New York: Plenum Press; Furnham, Adrian. "Belief in a just world: Research progress over the past decade." *Personality and individual differences* 34.5 (2003): 795–817; Oliver, Luis E. "Effects of a child's death on the marital relationship: A review." *OMEGA-Journal of Death and Dying* 39.3 (1999): 197–227; Hafer, Carolyn L., and Robbie Sutton. "Belief in a just world." *Handbook of social justice theory and research*. New York, NY: Springer New York, 2016. 145–160.
8. Albuquerque, Sara, Marco Pereira, and Isabel Narciso. "Couple's relationship after the death of a child: A systematic review." *Journal of Child and Family Studies* 25.1 (2016): 30–53; Dyregrov, Atle, Rolf Gjestad, and Kari Dyregrov. "Parental relationships following the loss of a child." *Journal of Loss and Trauma* 25.3 (2020): 224–244.
9. Helene J. Moriarty, Ruth Carroll, and Margaret Cotroneo, "Differences in Bereavement Reactions Within Couples Following Death of a Child," *Research in Nursing and Health* 19, no. 6 (1996): 461–69; Albuquerque et al., "Couple's Relationship After the Death of a Child."
10. C. Sue Carter, "Love as Embodied Medicine," *International Body Psychotherapy Journal* 18, no. 1 (2019): 19–25; C. Sue Carter and Stephen W. Porges, "The Biochemistry of Love: An Oxytocin Hypothesis," *EMBO Reports* 14, no. 1 (2013): 12–16; C. Sue Carter, "Oxytocin and Love: Myths, Metaphors, and Mysteries," *Comprehensive Psychoneuroendocrinology* 9 (2022): 100107; C. Sue Carter, "Oxytocin and the Healing Power of Love," in *The Handbook of Trauma-Transformative Practice: Emerging Therapeutic Frameworks for Supporting Individuals, Families or Communities Impacted by Abuse and Violence,* ed. Joe Tucci, Janise Mitchell, Stephen W. Porges, and Edward Tronick (2024), 96–106; Jean-Philippe Gouin, C. Sue Carter, Hossein Pournajafi-Nazarloo, Ronald Glaser, William B Malarkey, Timothy J. Loving, Jeffrey Stowell, and Janice K. Kiecolt-Glaser, "Marital Behavior, Oxytocin, Vasopressin, and Wound Healing," *Psychoneuroendocrinology* 35 (2010): 1082–90; C. Sue Carter, William M. Kenkel, Emily L. MacLean, Stephen R. Wilson, Allison M. Perkeybile, Justin R. Yee, and Michael A. Kingsbury, "Is Oxytocin 'Nature's Medicine'?," *Pharmacological Reviews* 72, no. 4 (2020): 829–61; Kiecolt-Glaser and Wilson, "Lovesick"; Janice K. Kiecolt-Glaser, Jean-Philippe Gouin, and Liisa Hantsoo, "Close Relationships, Inflammation, and Health," *Neuroscience and Biobehavioral*

Reviews 35, no. 1 (2010): 33–38; Bert N. Uchino, "Social Support and Health: A Review of Physiological Processes Potentially Underlying Links to Disease Outcomes," *Journal of Behavioral Medicine* 29 (2006): 377–87.

11. Amanda N. Gesselman, Silvia N. Bigatti, Justin R. Garcia, Kathryn Coe, David Cella, and Victoria L. Champion, "Spirituality, Emotional Distress, and Post-Traumatic Growth in Breast Cancer Survivors and Their Partners: An Actor–Partner Interdependence Modeling Approach," *Psycho-Oncology* 26, no. 10 (2017): 1691–99; Amanda N. Gesselman, Rebecca Ryan, William L. Yarber, Karen B. Vanterpool, Karly A. Beavers, Heather Francis, Brandon T. Grant, Katherine Wood, Cynthia A. Graham, Robin Milhausen, Stephanie A. Sanders, and Richard A. Crosby, "An Exploratory Test of a Couples-Based Condom-Use Intervention Designed to Promote Pleasurable and Safer Penile-Vaginal Sex Among University Students," *Journal of American College Health* 70, no. 6 (2022): 1665–72; Margaret Bennett-Brown, Amanda N. Gesselman, and Wendy R. Miller, "Constructive Communication Patterns as Related to Relationship Satisfaction, Seizure Severity, and Patient Activation Among People with Epilepsy," *Epilepsy and Behavior* 138 (2023): 108957; Amanda N. Gesselman, Rachel K. Wion, Justin R. Garcia, and Wendy R. Miller, "Relationship and Sexual Satisfaction Are Associated with Better Disease Self-Management in Persons with Epilepsy," *Epilepsy and Behavior* 119 (2021): 107937; Wendy R. Miller, Amanda N. Gesselman, Justin R. Garcia, Doyle Groves, and Janice M. Buelow, "Epilepsy-Related Romantic and Sexual Relationship Problems and Concerns: Indications from Internet Message Boards," *Epilepsy and Behavior* 74 (2017): 149–53.

12. Ulrika Fallbjörk, Berit H. Rasmussen, Sara Karlsson, and Pär Salander, "Aspects of Body Image After Mastectomy Due to Breast Cancer—A Two-Year Follow-Up Study," *European Journal of Oncology Nursing* 17, no. 3 (2013): 340–45; Beatriz Martins Faria, Isadora Martins Rodrigues, Luiza Verri Marquez, Úrsula da Silva Pires, and Sônia Vilges de Oliveira, "The Impact of Mastectomy on Body Image and Sexuality in Women with Breast Cancer: A Systematic Review," *Psicooncología* 18, no. 1 (2021): 91–115; Sevgi Koçan and Ayla Gürsoy, "Body Image of Women with Breast Cancer After Mastectomy: A Qualitative Research," *Journal of Breast Health* 12, no. 4 (2016): 145–50.

13. Ilaria Durosini, Stefano Triberti, Laura Savioni, Veronica Sebri, and Gabriella Pravettoni, "The Role of Emotion-Related Abilities in the Quality of Life of Breast Cancer Survivors: A Systematic Review," *International Journal of Environmental Research and Public Health* 19, no. 19 (2022): 12704; M. Tish Knobf, "Psychosocial Responses in Breast Cancer Survivors," *Seminars in Oncology Nursing* 23, no. 1 (2007): 71–83; Parisa Mokhatri-Hesari and Ali Montazeri, "Health-Related Quality of Life in Breast Cancer Patients: Review of Reviews from 2008 to 2018," *Health and Quality of Life Outcomes* 18 (2020): 1–25.

14. Elizabeth Rowland and Alison Metcalfe, "A Systematic Review of Men's Experiences of Their Partner's Mastectomy: Coping with Altered Bodies," *Psycho-Oncology* 23, no. 9 (2014): 963–74; Katerina Chronopoulou, Dimitrios Sakkas, and Dimitrios Damigos, "Caregiving Burden and Psychological Distress of Breast Cancer Patients' Husbands After Mastectomy," *International Journal of Caring Sciences* 9, no. 3 (2016): 909–13.

15. Alessandro Serretti and Alessandra Chiesa, "Sexual Side Effects of Pharmacological Treatment of Psychiatric Diseases," *Clinical Pharmacology and Therapeutics* 89, no. 1

(2011): 142–47; Sophie Marks, "A Clinical Review of Antidepressants, Their Sexual Side-Effects, Post-SSRI Sexual Dysfunction, and Serotonin Syndrome," *British Journal of Nursing* 32, no. 14 (2023): 678–82; Giovanni Corona, Michele Gacci, Enrico Baldi, Raffaele Mancina, Gianni Forti, and Mario Maggi, "Androgen Deprivation Therapy in Prostate Cancer: Focusing on Sexual Side Effects," *Journal of Sexual Medicine* 9, no. 3 (2012): 887–902; Dimitrios Terentes-Printzios, Nikolaos Ioakeimidis, Konstantinos Rokkas, and Charalambos Vlachopoulos, "Interactions Between Erectile Dysfunction, Cardiovascular Disease and Cardiovascular Drugs," *Nature Reviews Cardiology* 19, no. 1 (2022): 59–74; Lara J. Burrows, Maya Basha, and Andrew T. Goldstein, "The Effects of Hormonal Contraceptives on Female Sexuality: A Review," *Journal of Sexual Medicine* 9, no. 9 (2012): 2213–23.

16. Helen E. Fisher and James A. Thomson Jr., "Lust, Romance, Attachment: Do the Side Effects of Serotonin-Enhancing Antidepressants Jeopardize Romantic Love, Marriage, and Fertility?," in *Evolutionary Cognitive Neuroscience,* ed. Steven M. Platek, J. P. Keenan, and Todd K. Shackelford (Cambridge, MA: MIT Press, 2007), 245–83.
17. Vivian M. W. Ming, "Psychological Predictors of Marital Adjustment in Breast Cancer Patients," *Psychology, Health and Medicine* 7, no. 1 (2002): 37–51.
18. Ronald D. Adelman, Lyubov L. Tmanova, Diana Delgado, Sarah Dion, and Mark S. Lachs, "Caregiver Burden: A Clinical Review," *JAMA* 311, no. 10 (2014): 1052–60; Jung-won Lim and Brad Zebrack, "Caring for Family Members with Chronic Physical Illness: A Critical Review of Caregiver Literature," *Health and Quality of Life Outcomes* 2 (2004): 1–9; Q. P. Li, Y. W. Mak, and A. Y. Loke, "Spouses' Experience of Caregiving for Cancer Patients: A Literature Review," *International Nursing Review* 60, no. 2 (2013): 178–87; Recep Or and Asiye Kartal, "Influence of Caregiver Burden on Well-Being of Family Member Caregivers of Older Adults," *Psychogeriatrics* 19, no. 5 (2019): 482–90.
19. Roy F. Baumeister and Mark R. Leary, "The Need to Belong: Desire for Interpersonal Attachments as a Fundamental Human Motivation," *Psychological Bulletin* 117, no. 3 (1995): 497–529; C. Sue Carter, "Biological Perspectives on Social Attachment and Bonding," in *Attachment and Bonding: A New Synthesis,* ed. C. S. Carter, L. Ahnert, K. E. Grossmann, S. B. Hrdy, M. E. Lamb, S. W. Porges, and N. Sachser (Cambridge, MA: MIT Press, 2005), 85–100.
20. Julianne Holt-Lunstad, Timothy B. Smith, and J. Bradley Layton, "Social Relationships and Mortality Risk: A Meta-Analytic Review," *PLoS Medicine* 7, no. 7 (2010): e1000316; Julianne Holt-Lunstad, Timothy B. Smith, Mark Baker, Tyler Harris, and David Stephenson, "Loneliness and Social Isolation as Risk Factors for Mortality: A Meta-Analytic Review," *Perspectives on Psychological Science* 10, no. 2 (2015): 227–37; Julianne Holt-Lunstad, Theodore F. Robles, and David A. Sbarra, "Advancing Social Connection as a Public Health Priority in the United States," *American Psychologist* 72, no. 6 (2017): 517; Julianne Holt-Lunstad, "Social Connection as a Public Health Issue: The Evidence and a Systemic Framework for Prioritizing the 'Social' in Social Determinants of Health," *Annual Review of Public Health* 43, no. 1 (2022): 193–213; Julianne Holt-Lunstad, "Why Social Relationships Are Important for Physical Health: A Systems Approach to Understanding and Modifying Risk and Protection," *Annual Review of Psychology* 69, no. 1 (2018): 437–58.

21. Holt-Lunstad et al., "Social Relationships and Mortality Risk."
22. Nicole M. Cameron and Justin R. Garcia, "Maternal Effect and Offspring Development," in *Evolution's Empress: Darwinian Perspectives on the Nature of Women*, ed. Maryanne L. Fisher, Justin R. Garcia, and Rosemarie Sokol Chang (New York: Oxford University Press, 2013), 133–50.
23. Brooke C. Feeney and Nancy L. Collins, "A New Look at Social Support: A Theoretical Perspective on Thriving Through Relationships," *Personality and Social Psychology Review* 19, no. 2 (2015): 113–47; Mason G. Haber, Jay L. Cohen, Todd Lucas, and Boris B. Baltes, "The Relationship Between Self-Reported Received and Perceived Social Support: A Meta-Analytic Review," *American Journal of Community Psychology* 39 (2007): 133–44; Jingyi Wang, Farhana Mann, Brynmor Lloyd-Evans, Ruimin Ma, and Sonia Johnson, "Associations Between Loneliness and Perceived Social Support and Outcomes of Mental Health Problems: A Systematic Review," *BMC Psychiatry* 18 (2018): 1–16.
24. Patricia Churchland, *Conscience: The Origins of Moral Intuition* (New York: W. W. Norton, 2019); Shelley E. Taylor, *The Tending Instinct: How Nurturing Is Essential to Who We Are and How We Live* (New York: Macmillan, 2002); Brian Hare and Vanessa Woods, *Survival of the Friendliest: Understanding Our Origins and Rediscovering Our Common Humanity* (New York: Random House, 2020); Stephen E. Kessler, Timothy R. Bonnell, John M. Setchell, and Christopher A. Chapman, "Social Structure Facilitated the Evolution of Care-Giving as a Strategy for Disease Control in the Human Lineage," *Scientific Reports* 8, no. 1 (2018): 13997; Jean Decety, "The Neuroevolution of Empathy and Caring for Others: Why It Matters for Morality," in *New Frontiers in Social Neuroscience,* ed. Jean Decety et al. (Cham, Switzerland: Springer, 2013), 127–51; Frans B. M. de Waal, "Putting the Altruism Back into Altruism: The Evolution of Empathy," *Annual Review of Psychology* 59, no. 1 (2008): 279–300; Paul Gilbert, "The Evolution of Pro-Social Behavior: From Caring to Compassion," in *The Cambridge Handbook of Evolutionary Perspectives on Human Behavior,* ed. Louise Workman, William Reader, and Jerome H. Barkow (Cambridge: Cambridge University Press, 2020), 419–35; Jean Decety, "The Neuroevolution of Empathy," *Annals of the New York Academy of Sciences* 1231, no. 1 (2011): 35–45; Jean Decety, Gregory J. Norman, Gary G. Berntson, and John T. Cacioppo, "A Neurobehavioral Evolutionary Perspective on the Mechanisms Underlying Empathy," *Progress in Neurobiology* 98, no. 1 (2012): 38–48; Hillard S. Kaplan, Edward Schniter, Vernon L. Smith, and Barry J. Wilson, "Risk and the Evolution of Human Exchange," *Proceedings of the Royal Society B: Biological Sciences* 279, no. 1740 (2012): 2930–35; Robert Boyd and Peter J. Richerson, "Culture and the Evolution of Human Cooperation," *Philosophical Transactions of the Royal Society B: Biological Sciences* 364, no. 1533 (2009): 3281–88; Christopher L. Apicella and John B. Silk, "The Evolution of Human Cooperation," *Current Biology* 29, no. 11 (2019): R447–50.
25. Shelley E. Taylor, "Tend and Befriend: Biobehavioral Bases of Affiliation Under Stress," *Current Directions in Psychological Science* 15, no. 6 (2006): 273–77; Shelley E. Taylor, Lisa C. Klein, Barbara P. Lewis, Teresa L. Gruenewald, R. Andrew Gurung, and Jennifer A. Updegraff, "Biobehavioral Responses to Stress in Females: Tend-and-Befriend, Not Fight-or-Flight," *Psychological Review* 107, no. 3 (2000): 411–29; Shelley

E. Taylor, "Tend and Befriend Theory," in *Handbook of Theories of Social Psychology*, ed. Arie W. Kruglanski, E. Tory Higgins, and Agneta H. van Lange (Thousand Oaks, CA: Sage Publications, 2011).

26. Frans B. M. de Waal and Sarah D. Preston, "Mammalian Empathy: Behavioural Manifestations and Neural Basis," *Nature Reviews Neuroscience* 18, no. 8 (2017): 498–509; Frans de Waal, *The Age of Empathy: Nature's Lessons for a Kinder Society* (New York: Crown, 2010); de Waal, "Putting the Altruism Back into Altruism."

27. Marc Bekoff, *The Emotional Lives of Animals: A Leading Scientist Explores Animal Joy, Sorrow, and Empathy — and Why They Matter*, rev. ed (Novato, CA: New World Library, 2024).

28. Shifra Goldenberg, "Rare Video Shows Elephants 'Mourning' Matriarch's Death," *National Geographic*, December 9, 2021.

29. Barbara J. King, "When Animals Mourn," *Scientific American*, 309, no. 1 (2013): 62–67; Barbara J. King, *How Animals Grieve* (Chicago: University of Chicago Press, 2013).

30. Joshua M. Tybur and Debra Lieberman, "Human Pathogen Avoidance Adaptations," *Current Opinion in Psychology* 7 (2016): 6–11; Joshua M. Ackerman, "Disease Avoidance Hypothesis," in *Encyclopedia of Evolutionary Psychological Science*, ed. Todd K. Shackelford and Viviana A. Weekes-Shackelford (Cham, Switzerland: Springer, 2021), 2044–50; Valerie A. Curtis, "Infection-Avoidance Behaviour in Humans and Other Animals," *Trends in Immunology* 35, no. 10 (2014): 457–64; Benjamin Oosterhoff, Natalie J. Shook, and Ravi Iyer, "Disease Avoidance and Personality: A Meta-Analysis," *Journal of Research in Personality* 77 (2018): 47–56; Megan Oaten, Richard J. Stevenson, and Trevor I. Case, "Disease Avoidance as a Functional Basis for Stigmatization," *Philosophical Transactions of the Royal Society B: Biological Sciences* 366, no. 1583 (2011): 3433–52.

31. Hanna Kokko and Michael D. Jennions, "Parental Investment, Sexual Selection and Sex Ratios," *Journal of Evolutionary Biology* 21, no. 4 (2008): 919–48; E. Curio, "Animal Decision-Making and the 'Concorde Fallacy,'" *Trends in Ecology and Evolution* 2, no. 6 (1987): 148–52.

Chapter Ten: Love Again

1. Susan Lappan and Danielle Whittaker, eds., *The Gibbons: New Perspectives on Small Ape Socioecology and Population Biology* (New York: Springer, 2009).

2. Hironori Koda, "Gibbon Songs: Understanding the Evolution and Development of This Unique Form of Vocal Communication," in *Evolution of Gibbons and Siamang: Phylogeny, Morphology, and Cognition*, ed. Ulrich H. Reichard, Hirohisa Hirai, and Claudia Barelli (Cham, Switzerland: Springer, 2016), 349–59; Thomas Geissmann, "Duet-Splitting and the Evolution of Gibbon Songs," *Biological Reviews* 77, no. 1 (2002): 57–76; Esther Clarke, Ulrich H. Reichard, and Klaus Zuberbühler, "The Syntax and Meaning of Wild Gibbon Songs," *PLOS One* 1, no. 1 (2006): e73.

3. Daniel Feiler and Johannes Müller-Trede, "The One That Got Away: Overestimation of Forgone Alternatives as a Hidden Source of Regret," *Psychological Science* 33, no. 2 (2022): 314–24.

4. Esther Perel, *Mating in Captivity: Reconciling the Erotic and the Domestic* (New York: HarperCollins, 2006).

5. Ching-Teng L. Wang and Edward Schofer, "Coming Out of the Penumbras: World Culture and Cross-National Variation in Divorce Rates," *Social Forces* 97, no. 2 (2018): 675–704.
6. "Divorce Rates in the World: Divorce Rates by Country," Divorce.com, last updated July 15, 2024, https://divorce.com/blog/divorce-rates-in-the-world/.
7. Clayton Buck, Paul Hemez, and Lydia Anderson, "How Does Your State Compare with National Marriage and Divorce Trends?," US Census Bureau, October 8, 2024, https://www.census.gov/library/stories/2024/10/marriage-and-divorce.html; Grant Bailey, Lyman Stone, and Brad Wilcox, "Divorce in Decline: About 40% of Today's Marriages Will End in Divorce," Institute for Family Studies, July 24, 2025, https://ifstudies.org/blog/divorce-in-decline-about-40-of-todays-marriages-will-end-in-divorce.
8. Farzaneh Alizadeh, Abolfazi Mohammadbeigi, Reza Chaman, Fatemeh Kashefi, Ali Mohammad Nazari, and Zahra Motaghi, "Sexual and Reproductive Health Challenges in Temporary Marriage: A Systematic Review," *Journal of Research in Health Sciences* 21, no. 1 (2021): e00504; Yehezkel Margalit, "Temporary Marriage: A Comparison of the Jewish and Islamic Conceptions," *Journal of Law and Religion* 33, no. 1 (2018): 89–107.
9. Dawn O. Braithwaite and Leslie A. Baxter, "I Do Again: The Relational Dialectics of Renewing Marriage Vows," *Journal of Social and Personal Relationships* 12, no. 2 (1995): 177–98.
10. Lindsay R. Miller, Justin R. Garcia, and Amanda N. Gesselman, "Dating and Sexualities Across the Life Course: The Interactive Effects of Aging and Gender," *Journal of Aging Studies* 57 (2021): 100921; Peter B. Gray and Justin R. Garcia, "Aging and Human Sexual Behavior: Biocultural Perspectives," *Gerontology* 58, no. 5 (2012): 446–52; Cynthia A. Graham, Aleksandar Štulhofer, Theis Lange, Gert Martin Hald, Ana A. Carvalheira, Paul Enzlin, and Bente Træen, "Prevalence and Predictors of Sexual Difficulties and Associated Distress Among Partnered, Sexually Active Older Women in Norway, Denmark, Belgium, and Portugal," *Archives of Sexual Behavior* 49 (2020): 2951–61; Lauren B. Towler, Cynthia A. Graham, Felicity L. Bishop, and Sharron Hinchliff, "Sex and Relationships in Later Life: Older Adults' Experiences and Perceptions of Sexual Changes," *Journal of Sex Research* 60, no. 9 (2023): 1318–31; Osmo Kontula and Elina Haavio-Mannila, "The Impact of Aging on Human Sexual Activity and Sexual Desire," *Journal of Sex Research* 46, no. 1 (2009): 46–56.
11. Lisa L. Agate, Jolene M. Mullins, Ellas S. Prudent, and Thomas M. Liberti, "Strategies for Reaching Retirement Communities and Aging Social Networks: HIV/AIDS Prevention Activities Among Seniors in South Florida," *Journal of Acquired Immune Deficiency Syndromes* 33 (2003): S238–42; Osmo Kontula, "Sexuality Among Older Adults," in *International Handbook on the Demography of Sexuality*, ed. Amanda K. Bamle (Dordrecht: Springer, 2013), 195–214.
12. Gray and Garcia, "Aging and Human Sexual Behavior."
13. Stacy Tessler Lindau, L. Philip Schumm, Edward O. Laumann, Wendy Levinson, Colm A. O'Muircheartaigh, and Linda J. Waite, "A Study of Sexuality and Health Among Older Adults in the United States," *New England Journal of Medicine* 357, no. 8 (2007): 762–74.
14. Miller et al., "Dating and Sexualities Across the Life Course."

15. Trond Viggo Grøntvedt, Leif Edward Ottesen Kennair, and Mons Bendixen, "How Intercourse Frequency Is Affected by Relationship Length, Relationship Quality, and Sexual Strategies Using Couple Data," *Evolutionary Behavioral Sciences* 14, no. 2 (2020): 147–59; James K. McNulty, Carolyn A. Wenner, and Terri D. Fisher, "Longitudinal Associations Among Relationship Satisfaction, Sexual Satisfaction, and Frequency of Sex in Early Marriage," *Archives of Sexual Behavior* 45 (2016): 85–97.
16. Amanda N. Gesselman, Margaret Bennett-Brown, Simon Dubé, Ellen M. Kaufman, Jessica T. Campbell, and Justin R. Garcia, "The Lifelong Orgasm Gap: Exploring Age's Impact on Orgasm Rates," *Sexual Medicine* 12, no. 3 (2024): qfae042.
17. Peter B. Gray, Justin R. Garcia, and Amanda N. Gesselman, "Age-Related Patterns in Sexual Behaviors and Attitudes Among Single U.S. Adults: An Evolutionary Approach," *Evolutionary Behavioral Sciences* 13, no. 2 (2019): 111–26.
18. Gray and Garcia, "Aging and Human Sexual Behavior."
19. Arianna Romano, Silvia Bortolotti, Wilhelm Hofmann, Martin Praxmarer, and Matthias Sutter, "Generosity and Cooperation Across the Life Span: A Lab-in-the-Field Study," *Psychology and Aging* 36, no. 1 (2021): 108–18; Katherine I. Hunter and Matthew W. Linn, "Psychosocial Differences Between Elderly Volunteers and Non-Volunteers," *International Journal of Aging and Human Development* 12, no. 3 (1981): 205–13; Melissa M. Filkowski, Rebecca N. Cochran, and Brian W. Haas, "Altruistic Behavior: Mapping Responses in the Brain," *Neuroscience and Neuroeconomics* (2016): 65–75; Marc A. Musick and John Wilson, "Volunteering and Depression: The Role of Psychological and Social Resources in Different Age Groups," *Social Science and Medicine* 56, no. 2 (2003): 259–69.
20. Zane Dwyer, Nicholas Hookway, and Brendan Robards, "Navigating 'Thin' Dating Markets: Mid-Life Repartnering in the Era of Dating Apps and Websites," *Journal of Sociology* 57, no. 3 (2020): 647–63; Sarah McWilliams and Anne E. Barrett, "Online Dating in Middle and Later Life: Gendered Expectations and Experiences," *Journal of Family Issues* 35, no. 3 (2014): 411–36; Miller et al., "Dating and Sexualities Across the Life Course: The Interactive Effects of Aging and Gender," *Journal of Aging Studies* 57 (2021): 100921.
21. Justin J. Lehmiller and Christopher R. Agnew, "May-December Paradoxes: An Exploration of Age-Gap Relationships in Western Society," in *The Dark Side of Close Relationships II*, ed. William R. Cupach and Brian H. Spritzerg (New York: Routledge, 2011), 39–61; Brian Collisson and L. P. De Leon, "Perceived Inequity Predicts Prejudice Towards Age-Gap Relationships," *Current Psychology* 39, no. 6 (2020): 2108–15; Angela Niccolai and Melissa Swauger, "Minding the (Age) Gap: The Identity and Emotion Work of Men and Women in Age-Discrepant Romantic Relationships," *Sociological Focus* 54, no. 1 (2021): 19–38; Justin J. Lehmiller and Ryan A. Christopher, "Commitment in Age-Gap Heterosexual Romantic Relationships: A Test of Evolutionary and Socio-Cultural Predictions," *Psychology of Women Quarterly* 32, no. 1 (2008): 74–82.
22. Gray et al., "Age-Related Patterns in Sexual Behaviors and Attitudes Among Single U.S. Adults"; Gray and Garcia, "Aging and Human Sexual Behavior"; Peter B. Gray and Justin R. Garcia, *Evolution and Human Sexual Behavior* (Cambridge, MA: Harvard University Press, 2013).

23. Koda, "Gibbon Songs"; Peng Han, Hong Gao Ma, Zhi Dong Wang, Ping Ling Fan, and Peng Fei Fan, "Vocal Differences in Note, Sequence and Great Call Sequence Among Three Closely Related *Nomascus* Gibbon Species," *International Journal of Primatology* 46, no. 1 (2025): 45–68.

Conclusion: Living and Dying for Love

1. Solve OCO, "Love in the Time of AI Girlfriends," Medium, accessed April 23, 2025, https://solveoco.medium.com/love-in-the-time-of-ai-girlfriends-643fc08c64e4#:~:text=The%20demand%20for%20AI%20girlfriends,themselves%2C%20attracting%20thousands%20of%20users; Chris Westfall, "As AI Usage Increases at Work, Searches for 'AI Girlfriend' up 2400%," *Forbes*, September 29, 2023.

INDEX

Acevedo, Bianca, 23, 162
addiction, love as, 159
affairs. See infidelity
affiliative behavior, 26–28
African elephants, 137
aging, love and, 187–192
Agnew, Chris, 165
agricultural revolution, impact on courtship, 37–42
airport dating app strategy, 59–61
Aka, 47
alloparents, 35–36, 102, 122
American Hookup: The New Culture of Sex on Campus (Wade), 58
ancestress hypothesis, 121–122
Anderson, Kermyt, 145
animal courtship displays, 51–52
argonaut octopus, 98
Aron, Arthur, 75, 77, 97
arousal, misattribution of, 74–77
arranged marriages, 40–41
artificial intelligence, 198–199

Baumeister, Roy, 64
behavioral immune system, 99–100
behavioral synchronization, 53–56
betrayal, 151, 153–154. See also infidelity
bluebirds, 144
boomerang generation, 123–124
brain
    bonding in, 32–34
    impact of breakups on, 159–160, 162
    intimacy and, 23
    long-term love in, 114

breakups
    checking on exes after, 161–163
    complexities of, 161–166
    divorce, 129, 137–138, 163, 186
    gendered patterns, 163–164
    grief after, 155–158
    impact on brain, 159–160
    learning from past relationships, 187
    living together after, 156–157
    reasons for, 164–166
    temporary breaks, 160
    ways to get over, 166–167
breastfeeding, 101–102
Bruch, Elizabeth, 61
Buss, David, 47–48, 149

cancer. See illness
caregiver burden, 176
caretaking
    as hallmark of humanity, 178–180
    illness and sex, 173–176
    impact on relationship, 172–173
    navigating unpredictable illness, 169–173
    power of care, 180–181
    role in evolution of species, 177–178
    in sickness and in health, 176–180
    social support and, 176–177
Carter, Sue, 32–34, 173
casual sex
    after breakups, 166–167
    rise of hookup culture, 56–59, 85–86
catfishing, 89–90
Catron, Mandy Len, 77

Index

character traits
  caretaking, 178
  dealbreakers, 70–71
  preferences in, 65–66
cheating. See infidelity
children
  alloparents, 35–36, 102, 122
  boomerang generation, 123–124
  breastfeeding, 101–102
  death of, 172–173
  in evolution of pair bonding, 35–36
  fertility rates, 120
  genetic paternity, 145–147
  mate selection and, 99–100
  opinion of older parent's romantic relationships, 191–192
  outcomes in diverse family structures, 122–123
  pair bonding needed for, 120–121
  parental investment theory, 68–69
  parenting and sexual satisfaction, 130–132
  partible paternity, 110
chimpanzees, 178
coaches, dating, 82–83
Coe, Kathryn, 121–122
cohabitation. See also nesting
  benefits of, 129–130
  after breakup, 156–157
  changes in social norms about, 19
  trends in, 117–119
coital frequency, 84–85
communication
  accelerating closeness, 77–81
  setting rules in nonmonogamy, 150–152
  and sexual satisfaction, 127
community, role in courtship, 43–44
companionate love, 124–127, 190
compassion for recently broken-up, 158
Concorde fallacy, 179
Conroy-Beam, Daniel, 63
consensual nonmonogamy, 107–113, 150–152
Coolidge effect, 138–139
cooperative breeding, 35–36, 102
cost-benefit analyses, 64
"couch-and-cuddle" phase, 124–127

courtship
  accelerating closeness in, 77–81
  among older adults, 187–192
  catfishing, 89–90
  dance in, 53–56
  dating coaches, 82–83
  dealbreakers, 70–71
  displays of animals, 51–52
  economics of mating market, 61–63
  first sexual encounters, 83–85
  gaming dating apps, 59–61
  genes, 67–69
  hookup culture, rise of, 56–59
  impact of agricultural revolution on, 37–42
  impact of internet on, 42–47
  misattribution of arousal, 74–77
  modern concept of, 74
  overview, 74
  role of genes in, 67–69
  sexting, 87–90
  showing your hand, 63–67
  signaling exclusivity, 90–93
  slow love, 84–87
  video dates, 81–82
COVID-19 pandemic, 21–24, 197
curiosity, 81

dance, 53–56
dating
  accelerating closeness, 77–81
  among older adults, 187–192
  catfishing, 89–90
  dance in, 53–56
  dating coaches, 82–83
  dealbreakers, 70–71
  economics of mating market, 61–63
  first sexual encounters, 83–85
  gaming dating apps, 59–61
  hookup culture, rise of, 56–59
  impact of agricultural revolution on, 37–42
  impact of internet on, 42–47
  misattribution of arousal, 74–77
  modern concept of, 74
  role of genes in, 67–69
  sexting, 87–90
  showing your hand, 63–67

Index

signaling exclusivity, 90–93
slow love, 84–87
video dates, 81–82
dating apps
  dating coaches, 82–83
  evolution of courtship, 42–47
  gaming geographical data, 59–61
  impact on intimacy, 24–28
  for older adults, 26, 187
dating coaches, 82–83
de Waal, Frans, 178
dealbreakers, 70–71
death
  of children, 172–173
  loving again after loss, 192–195
  of partner, 158
dick pics, 87–89
digital technology. See also dating apps
  dealing with breakups online, 161–162
  and future of intimacy, 198–199
  impact on courtship, 42–47
  impact on intimacy, 24–28
  intergenerational cultural transmission, 25–26
  signaling exclusivity with, 90–93
divorce, 129, 137–138, 163, 186
Dobbs decision, 140
dopamine genes, 140–146
Dutton, Donald, 75

economics of mating market, 61–63, 64
Eight Dates: Essential Conversations for a Lifetime of Love (Gottman & Gottman), 78
elephants, 137, 178
emotional infidelity, 149. See also infidelity
emotional vulnerability, 85–86
empathy, as factor in fidelity, 152
evolutionary history of pair bonding, 34–37
exclusivity, signaling, 90–93
exes, checking on, 161–163
extra-pair copulation, 153

f'a'afafine, 121–122
"Facebook official," 91
face-to-face sex, 106–107

falling in love, 48
family. See also children
  alloparents, 35–36, 102, 122
  death of children, 172–173
  diverse structures of, 122–123, 128–130
  evolutionary history of, 34–37
  fertility rates, 120
  genetic paternity, 145–147
  influence on relationships, 100–104
  nesting and, 121–124
  parental investment theory, 68–69
  partible paternity, 110
  sexual satisfaction and parenting, 130–132
fertility rates, 120
fight-or-flight response, 178
Finkel, Eli, 70
first dates, 73–76, 81–83
Fisher, Helen, 12, 25, 82, 84, 85
fraternal polyandry, 109–110
Frederick, David, 66, 125–126
Frost, Robert, 55

genetics
  factors in infidelity, 140–146
  genetic paternity, 145–147
  preferences based on reproduction, 67–69
genital images, texting of, 87–89
geographical settings on dating apps, 59–61
Gesselman, Amanda, 71, 82, 160, 174, 189
getting over breakups, 166–167
gibbons, 184–185, 192–194
Gillespie, Brian, 125–126
golden years, love in, 187–192
Gottman, John, 78
Gottman, Julie Schwartz, 78
Gray, Peter, 131, 189
grief
  among elephants, 178
  after breakups, 155–158
  after death of loved ones, 172–173, 177
  loving again after loss, 192–195

Hatfield, Elaine, 124–127
health
  benefits of love, 173–174
  impact of social support on, 177

Index

height of partners, 62
Holt-Lunstad, Julianne, 177
hookup culture, rise of, 56–59, 85–86, 147
household labor, 132
Hrdy, Sarah Blaffer, 35–36

illness
    impact on relationship, 172–173
    impact on sex, 173–176
    navigating unpredictable, 169–173
    power of, 180–181
    in sickness and in health, 176–180
    social support and, 176–177
Inclusion of the Other in the Self (IOS) scale, 97
infidelity
    biological, social, and genetic factors, 138–145
    consensual nonmonogamy and, 150–152
    genetic paternity and benefits, 145–147
    mate guarding, 135–138
    reasons for, 147–150
    reasons for fidelity, 152–154
    repairing relationship after, 153–154, 164
the infidelity gene, 140–146
intergenerational cultural transmission, 25–26
internet
    AI girlfriends, 198–199
    dealing with breakups on, 161–162
    impact on courtship, 42–47
    impact on intimacy and loneliness, 24–28
    intergenerational cultural transmission, 25–26
    signaling exclusivity, 90–93
intimacy
    COVID-19 pandemic and, 21–24
    defined, 4–5
    digital disconnect and, 24–28
    impact on life satisfaction, 20
    need for, 17–20, 49
    need for physical touch, 26–28
    neuroimaging studies on, 23
    pleasure and pair bonding, 47–49
    and sex drive, 5–7
intimacy crisis, 20

IOS (Inclusion of the Other in the Self) scale, 97

Japanese red-crowned cranes, 51
jealousy, 135–138
Johnson, Colin, 38–39
just-world fallacy, 172–173

kin influence, 100–104
kindness, as desired quality, 65–67
Kinsey, Alfred C., 9–10, 105
Kinsey Institute, 9–12
Kinsey Reports, 10
kissing, lack of chemistry in, 43

lactational aggression, 130
lesbian bed death, 126–127
Lever, Janet, 125–126
LGBTQ+
    child outcomes in same-gender parent households, 122–123
    consensual nonmonogamy among, 110
    mate preferences, 65, 67
    sustaining sexual passion in relationships, 126–127
life history theory, 119
Lindau, Stacy, 188
Lippa, Richard, 66
loneliness, 20, 21–24, 157
long-distance relationships, 79–80
long-term relationships. See also breakups; caretaking; nesting; pair bonding
    becoming mates, 95–97
    brain regions involved in, 113–115
    changes in social norms and, 18–20
    consensual nonmonogamy, 107–113, 150–152
    division of labor in, 132
    family influence, 100–104
    intimacy and sex drive paradox, 4–7
    investing in relationship, 97–100
    maintaining intimacy in, 79–81
    myths about, 10–11
    role of sex in bonding, 104–107
    sustaining sexual passion in, 124–127
    topics of arguments in, 156

Index

love
  as addiction, 159
  arranged marriages, 40–41
  as best medicine, 173–174
  brain and, 23, 32–34, 114
  consensual nonmonogamy and, 111–113
  evolution of courtship, 37–42
  evolutionary history of, 34–37
  falling in love, 29–30, 48
  impact of internet on, 42–47
  physical and emotional sensations of, 30
  on pleasure and pair bonds, 47–49
  role of oxytocin, 32–34
  slow love, 84–87
  study of, 11–12
loverese, 90–91
loving again
  ability to form new relationships, 185–187
  gibbon songs, 184–185, 192–194
  in golden years, 187–192
  learning from past relationships, 187
  loving differently after loss, 192–195
  multiple relationships in one, 185–187

magpies, 178
Maner, Jon, 114
Marcotte, Alexa, 88
marriage. *See also* caretaking; infidelity; nesting
  arranged, 40–41
  average length of, 186
  benefits of, 129–130
  consensual nonmonogamy, 107–113, 150–152
  divorce, 129, 137–138, 163, 186
  family roles, 121–124
  in good times and bad, 132–133
  impact of agricultural revolution on, 38–42
  maintaining intimacy in, 79–81
  nuclear family and other structures, 128–130
  sexual satisfaction and parenting, 130–132
  slow love, 84–87
  suffocation model, 70
  sustaining sexual passion, 124–127
  time-limited, 186
  topics of arguments in, 156
mastectomy, impact on sex life, 174–175
Match, 12, 25
mate choice
  becoming mates, 95–97
  brain regions involved in, 113–115
  consensual nonmonogamy, 107–113
  family influence, 100–104
  investing in relationship, 97–100
  role of sex in bonding, 104–107
mate guarding, 135–138
mate values, 62–63, 65–66, 70
mating market, economics of, 61–63, 64
medications, impact on sex life, 175, 188
men
  consensual nonmonogamy among, 107–111
  impact of breakups on, 163–164
  infidelity of, 139–140, 149
  orgasm in, 33
  sex among older adults, 187–192
  sexual satisfaction and parenting, 130–132
Meston, Cindy, 48
Miller, Lisa, 188
misattribution of arousal, 74–77
montane voles, 32
Moors, Amy, 110
Mpimbwe people, 123
mutual rituals, 80–81
myths about relationships, 10–11

Neolithic revolution, 38
nesting
  building nests, 119–121
  cohabitation, 117–119
  division of labor in, 132
  family roles, 121–124
  going nuclear, 128–130
  in good times and bad, 132–133
  keeping flame alive, 124–127
  sexual satisfaction and parenting, 130–132
neuroimaging studies
  on bonding, 32–34
  on impact of breakups, 159–160, 162
  on intimacy, 23
  on long-term love, 114

Index

Ngandu, 47
The Normal Bar (Northrup, Schwartz, & Witte), 90–91
Northrup, Chrisanna, 90–91
nuclear family, 128–130
nude images, texting of, 87–89

older adults, love among, 187–192
online dating
    AI girlfriends, 198–199
    catfishing, 89–90
    dating coaches, 82–83
    evolution of courtship, 42–47
    gaming geographical data, 59–61
    impact on intimacy, 24–28
    for older adults, 26, 187
    signaling exclusivity, 90–93
    unsolicited nude images, 87–89
open relationships, 107–113, 150–152
oral sex, 83–84, 127, 189
orgasm, 33, 47, 105–106, 127
OurTime, 26, 187
oxytocin, 32–34

pair bonding. See also breakups; caretaking; nesting
    becoming mates, 95–97
    brain chemistry in, 32–34
    brain regions involved in, 113–115
    consensual nonmonogamy, 107–113, 150–152
    evolution of, 34–37
    family influence, 100–104
    in gibbons, 184–185, 192–194
    impact of agricultural revolution on, 37–42
    investing in relationship, 97–100
    overview, 18
    sex and, 47–49, 104–107
    sustaining sexual passion in, 124–127
paradox of choice, 60
parenting
    alloparents, 35–36, 102, 122
    death of children, 172–173
    parental investment theory, 68–69
    partible paternity, 110
    sexual satisfaction and, 130–132
partner preferences, 61–63, 65–66, 70

passion, keeping alive, 124–127
paternity, genetic, 145–147
Pauli, Gabriele, 186
peacock spider, 51–52
peacocking, 63–67
Perel, Esther, 186
pet names, 90–91
pheromones, 43
physical contact
    need for, 26–28
    oxytocin and, 32–34
    physical connection in dance, 52–56
polyamory, 107–113
polygamy, 107–113
polyvagal theory, 76
Porges, Stephen, 76
positive illusions, 113–114
prairie voles, 32
preferences, partner, 61–63, 65–66, 70
preferential sociality, 33

questions for intimate partners, 77–81

red-winged blackbirds, 153
relationship dissolution. See breakups
relationship medicine, 180
reproduction
    fertility rates, 120
    mate selection and, 99–100
    preferences based on, 68–69
    sexual satisfaction and parenting, 130–132
risk-taking
    casual sex, 85–86
    the infidelity gene, 140–146
rituals, mutual, 80–81
robots, 199
romantic baby talk, 90–91

sad music, listening to after breakups, 166
sage grouse, 52
Schmitt, David, 139
Schwartz, Pepper, 90–91
Sear, Rebecca, 120, 128
self-disclosure, 77–81
self-expansion, mate preferences regarding, 67
Selterman, Dylan, 147–148
sensation-seeking genes, 140–146

Index

sensory information, role in courtship, 42–46
sex. *See also* infidelity
   bonding through, 104–107
   after breakups, 166–167
   consensual nonmonogamy, 107–113, 150–152
   COVID-19 pandemic and, 22
   evolution of pair bonding, 34–37
   first encounters when dating, 83–84
   frequency of, 84–85
   hookup culture, rise of, 56–59, 85–86
   impact of illness on, 173–176
   keeping flame alive, 124–127
   in older adults, 187–192
   oral, 83–84, 127, 189
   orgasm, 33, 47, 105–106
   pair bonding and pleasure, 47–49
   and parenting, 130–132
   sex drive, 5–7
   sex interview, 85
   sex recession, 28, 84–85
   slow love and, 84–87
sexting, 87–89
sexual monogamy. *See also* infidelity
   marital arrangements and, 41–42
   mate guarding, 137
   versus social monogamy, 37
shiva, 158
showing your hand, 63–67
sickness. *See* illness
signaling exclusivity, 90–93
singlehood, rise in, 18–20
Singles in America study, 12–13, 25, 76, 107, 131, 189
slow love, 84–87, 98
smell, sense of, 42–43
social media
   dealing with breakups on, 161–162
   signaling exclusivity, 90–93
social monogamy
   impact of agricultural revolution on, 37–42
   impact of internet on, 42–47
   mate guarding, 136–137
   versus sexual monogamy, 37
social support, importance of, 176–177
soulmates, myth of, 185
starling murmuration, 53

suffocation model, 70
super synchronizers, 55
swingers, 107–108

taking a break in relationship, 160
Taylor, Shelly, 178
tend-and-befriend response, 178
testosterone, impact of parenthood on, 131
thirty-six questions for cultivating intimacy, 77–78
threesomes, 111
time-limited marriages, 186
timing of relationships, 165
"To Fall in Love with Anyone, Do This" (Catron), 77
Todd, Peter, 68–69
touch (physical)
   need for, 26–28
   oxytocin and, 32–34
trust, 11
TV shows, watching together, 79–80

Vasey, Paul, 122
video dates, 81–82
virginity, 71
Vitzthum, Virginia, 65
vulnerability, 85–86, 167

Wade, Lisa, 58
weedy sea dragons, 51
western gulls, 145
White Whale, 3–4
Witte, James, 90–91
women
   ancestress hypothesis, 121–122
   breastfeeding, 101–102
   consensual nonmonogamy among, 107–111
   fertility rates, 120
   impact of breakups on, 163–164
   infidelity of, 139–140, 149
   orgasm in, 33
   sex among older adults, 187–192
   sexual satisfaction and parenting, 130–132

zhiji, 190
zip-lining, 73–74